STUDENT UNIT GUIDE

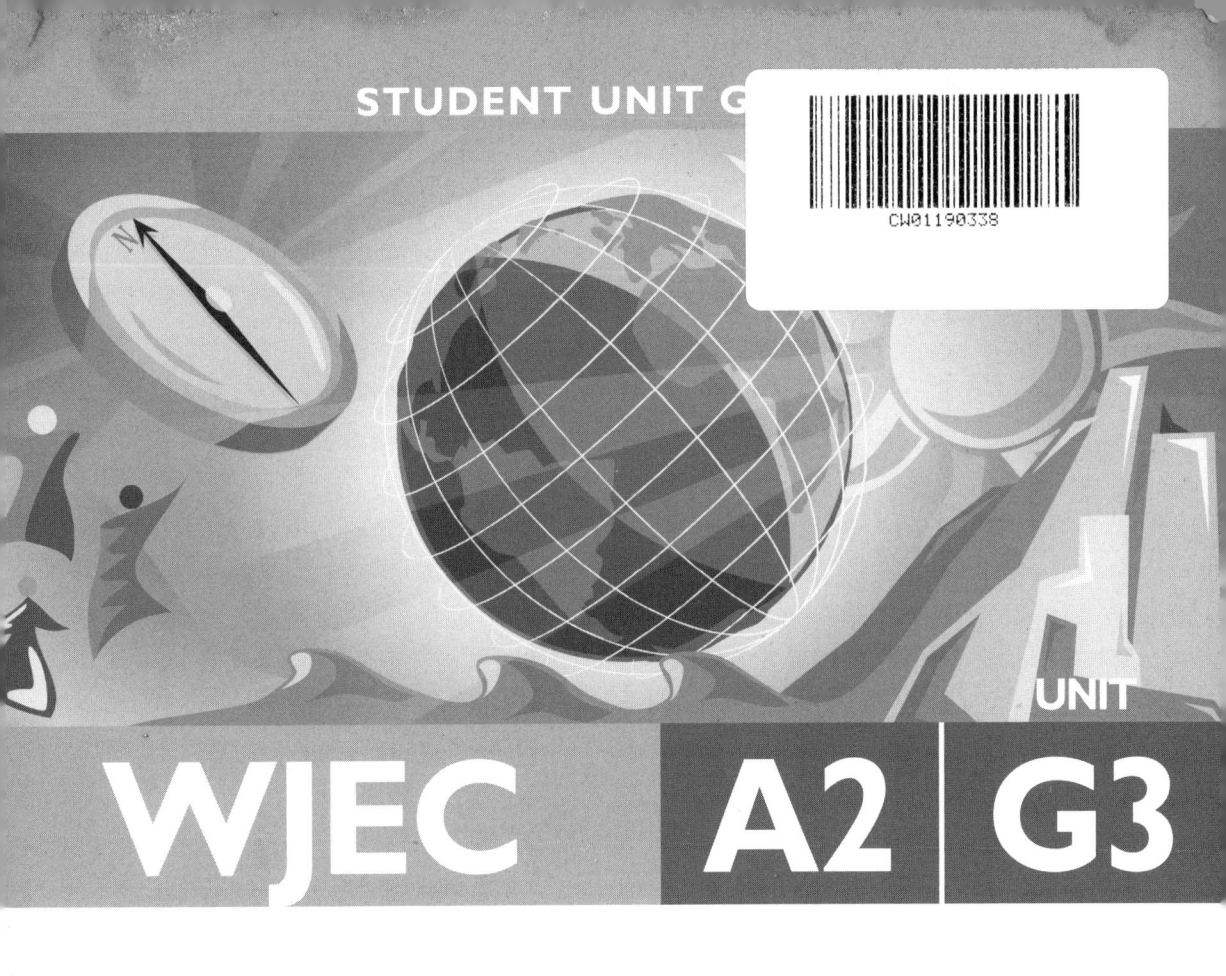

UNIT

WJEC A2 G3

Geography

Contemporary Themes and
Research in Geography

Nicky King and David Burtenshaw

With thanks to Gill Miller of Chester University for her contribution to Themes 6(a) and 6(b)

Philip Allan Updates, an imprint of Hodder Education, an Hachette UK company, Market Place, Deddington, Oxfordshire OX15 0SE

Orders
Bookpoint Ltd, 130 Milton Park, Abingdon, Oxfordshire OX14 4SB
tel: 01235 827720
fax: 01235 400454
e-mail: uk.orders@bookpoint.co.uk
Lines are open 9.00 a.m.–5.00 p.m., Monday to Saturday, with a 24-hour message answering service. You can also order through the Philip Allan Updates website: www.philipallan.co.uk

© Philip Allan Updates 2010

ISBN 978-1-4441-1084-5

First printed 2010
Impression number 5 4 3 2 1
Year 2014 2013 2012 2011 2010

All rights reserved; no part of this publication may be reproduced, stored in a retrieval system, or transmitted, in any other form or by any means, electronic, mechanical, photocopying, recording or otherwise without either the prior written permission of Philip Allan Updates or a licence permitting restricted copying in the United Kingdom issued by the Copyright Licensing Agency Ltd, Saffron House, 6–10 Kirby Street, London EC1N 8TS.

This guide has been written specifically to support students preparing for the WJEC A2 Geography Unit G3 examination. The content has been neither approved nor endorsed by WJEC and remains the sole responsibility of the authors.

Typeset by Philip Allan Updates

Printed by MPG Books, Bodmin

Hachette UK's policy is to use papers that are natural, renewable and recyclable products and made from wood grown in sustainable forests. The logging and manufacturing processes are expected to conform to the environmental regulations of the country of origin.

WJEC Unit 3

Contents

Introduction
About this guide... 4

Content Guidance
About this section 12

Section A Contemporary themes 13
Theme 1 Extreme environments 13
Theme 2(a) Glacial landforms and their management 17
Theme 2(b) Coastal landforms and their management 24
Theme 3 Climatic hazards 32
Theme 4 Development 40
Theme 5 Globalisation 47
Theme 6(a) Emerging Asia: China 52
Theme 6(b) Emerging Asia: India 58

Section B Individual research enquiry 66

Questions & Answers
About this section 76

Section A Contemporary themes
Q1 Theme 3 Climatic hazards 77
Q2 Theme 5 Globalisation 80
Q3 Theme 2(b) Coastal landforms and their management 83
Q4 Theme 4 Development 85

Section B Individual research enquiry
Q5 Environmental psychology 87
Q6 Rivers 89
Q7 Geography of disease 90
Q8 Leisure and recreation 92

Introduction

About this guide

The purpose of this guide is to help you understand what is required to do well in **Unit G3: Contemporary Themes and Research in Geography**. The guide is divided into three sections.

This **Introduction** explains the structure of the guide and the nature of the Unit G3 component of the A2 level qualification. It also provides some general advice on how to approach the unit examination.

The **Content Guidance** section sets out the *bare essentials* of the specification for this unit. Diagrams are used to help your understanding; many are simple to draw and could be used in the exam.

The **Question & Answer** section includes seven sample exam questions in the style of the unit examination. Sample answers of differing standards are provided, as well as examiner's comments on how to tackle each question and on where marks have been gained or lost in the sample answers.

Unit G3 examination

Timing

The examination lasts for a total of 2 hours 15 minutes and counts for 60% of your A2 qualification, but it is broken down into two parts:
- Section A: Contemporary Themes in Geography, lasts 1 hour 30 minutes.
- Section B: Research in Geography, lasts 45 minutes.

The two parts of the examination will be handed out separately in the examination room. *Section A answers will be collected in by the exam invigilators before you are given the separate paper for Section B.*

In Section A each question is worth 25 marks and you are required to answer two questions and divide up your time equally between them, giving you 45 minutes to answer each question. In Section B you will need to answer a compulsory two-part question from the research theme and topic area that you have selected. The part (a) question is a 10-mark generic question that examines your understanding of the enquiry approach, whereas the part (b) question is a 15-mark question that will examine a specific aspect of your own particular research topic. You will therefore need to spend 18 minutes (just under 20 minutes) on question (a) and 27 minutes (just over 25 minutes) on question (b).

Choice of questions

Section A
Section A is an essay paper, which is divided into two parts, each made up of eight questions. You must answer one question from Contemporary Physical Themes 1–3, and one question from Contemporary Human Themes 4–6. Each question is worth 25 marks or 40 UMS (total 80 UMS).

Section B
This part of the specification provides you with the opportunity to carry out individual research and out-of-classroom work, including fieldwork, on a pre-set topic on one research investigation theme listed on page 66.

You are required to answer a compulsory two-part question from the research theme and topic area that you have selected. It is possible that you will have carried out research on a different topic from others in your class/group as you will have been encouraged to research a topic that particularly interests you. However, your tutor may have given all of your class the same research topic. This compulsory two-part question is worth 25 marks or 40 UMS.

Synopticity, concepts, geographical terms and case studies

In addition to examining your knowledge and understanding of themes you have studied for the Unit G3 paper, the examination will test your ability to identify and analyse the connections between these themes and the relevant aspects of geography already studied as part of your AS Geography course. The ability to make connections between different parts of the course is very important to success at A2 level and is known as **synopticity**. For example, if you have studied the Physical Theme 3, Climatic Hazards, you should review your knowledge and understanding of Climatic Change that you will have studied for the Unit G1 component at AS level. In order to achieve the highest marks at A2, you need to demonstrate to the examiner that your knowledge of the contemporary theme you have studied is very sound and that you have a very firm grasp of the definitions and concepts that are central to the topics you have covered.

In addition, you need to acquire a sound knowledge of supporting, relevant **case studies** and the ability, where appropriate, to produce clear, relevant and accurate annotated diagrams that are effectively integrated into your answers. Using the correct vocabulary is also essential. It may be useful to compile a list of **key terms**: use textbooks, teacher's notes and your own class notes to build up a glossary of relevant geographical terms. As this is an A2 unit you will be expected to read around the contemporary theme that you are studying to give you a broader and better-informed knowledge base. Look for relevant articles in subject-specific magazines such as *Geography Review*, *Topic Eye* and *The Geographical*, as well as relevant geography textbooks and quality newspapers.

Mark scheme

Section A questions are marked on five levels.

Level	Description of quality	Marks range	What examiners are looking for
5	Very good	21–25	Very good knowledge and understanding used critically. An ability to evaluate arguments. Good, possibly original examples. A clear, coherent essay which is grammatically correct. Good diagrams and sketch maps where appropriate.
4	Good	16–20	Good knowledge and understanding with some critical awareness. Evaluation more patchy. A clearly structured essay that uses good English but argues points soundly rather than strongly. Appropriate diagrams and maps not always fully labelled.
3	Average	11–15	Knowledge and understanding present but some points may be partial and lack exemplar support. Mainly uses text or taught examples of variable quality. Language is straightforward and is in need of more complex use of geographical arguments
2	Marginal	6–10	Some knowledge and understanding but with gaps and misconceptions. The answer is not very broad in its cover and has only limited support from examples, diagrams and maps. Language is variable and slips occur.
1	Weak	1–5	Knowledge and understanding is very basic and lacking supporting evidence. Shows evidence of not understanding the question. Use of language of geography and written style contains slips.

A fuller, official copy of the generic mark schemes can be found in the Specimen Assessment Materials at your college or school.

Section B contains 10 two-part questions marked out of 10 (part a) and 15 (part b).

Part (a) is marked on three levels.

Part (a)	What examiners are looking for
Level 3 8–10 marks	Very good knowledge and understanding used critically which is applied to research route to enquiry. The work is obviously based on research and uses it to provide good supporting evidence. A clear, coherent mini-essay which is grammatically correct.
Level 2 4–7 marks	Good knowledge and understanding with some critical awareness of the route to enquiry. A clearly structured mini-essay that uses good English but argues points soundly rather than strongly. Appropriate diagrams and maps not always fully labelled.
Level 1 1–3 marks	Some limited knowledge and understanding of some aspects of the route to enquiry present but some points may be partial and lack exemplar support from the research. May use taught material of variable relevance. 'All I know' rather than an answer to the question. Language is variable, lacking paragraphs and may have weak grammar and syntax.

WJEC Unit 3

Part (b) is marked on four levels. The levels will partially relate to the expected content in your answers.

Part (b)	Characteristics of level (it is not necessary to meet all of the characteristics to be placed in a level)
Level 4 **13–15 marks**	Provides title of research. Very good knowledge of the topic studied and a critical awareness of the route to enquiry as applied to the topic in question. Provides very good support from own research. May have some good diagrammatic material and maps to support answer. Written in a sound, coherent essay style, which is grammatically correct, with a sequence of ideas that enables the question to be answered fully. Able to evaluate when required. Concludes in relation to the question.
Level 3 **9–12 marks**	Provides title of research. Good knowledge of the topic with some gaps. Understanding of the route to enquiry is present but may occasionally be unsound. Good knowledge and understanding with some critical awareness. Evaluation more patchy. A clearly structured essay that uses good English but argues points soundly rather than strongly. May have a conclusion.
Level 2 **5–8 marks**	Provides title of research. Knowledge and understanding present but some points may be partial and lack exemplar support from research theme studied. Verges on the formulaic answer which is possibly almost identical for all students in centre. Language is straightforward and will possibly lack paragraphing. Perhaps going off at a tangent with an 'all I know' answer. Tails off and may not have a conclusion.
Level 1 **1–4 marks**	Possibly neglects to state title. Some knowledge and understanding but with gaps and misconceptions that indicate an inability to understand the question. Evidence that the research was superficial. Only limited support from research. Language is variable and slips occur.

Examination skills

It is important at the outset to work out very clearly the requirements of the question you have decided to answer. It is rare that learnt material can simply be reproduced. It is important that you select only those parts of learnt material that are relevant to the question set. Avoid lengthy introductions that do not directly address the question and simply provide general background material. From the outset go straight to the point of the question. To ensure continuity and relevance and to avoid drifting off the point of the question, it is important to link each major point back to the question. In this respect, planning your response at the very beginning is really important. It is recommended that you spend a few minutes planning your answer at the outset, either using bullet points that identify the key content for each paragraph, or by using 'spider' diagrams.

Managing in the examination room

You will be issued with the examination paper for Section A and at the end of 1 hour 30 minutes that paper will collected and the Section B examination paper will be handed out. In effect it is two examination papers with a short gap. Here are some tips for good performance in the examination:
- Use a highlighter pen to emphasise the subject matter of your chosen questions.
- Use a different colour to highlight the command words.

- Always write a brief plan for each answer before you start. A plan that is not crossed out may be used by the marker in the case of an incomplete answer.
- Keep strictly to time.
- Keep to the formal essay format of sentences grouped into paragraphs arranged in a logical order. DO NOT use text-speak.
- Essays should have a brief introduction that identifies the direction of your argument and a brief conclusion that summarises your essay.
- Try to leave time to reread your essays and correct inaccuracies and spelling.
- Remember that geography at this level is about the complexity of causes, issues and problems because the interrelationships between people and environments are complex.
- You must be able to draw together information from a variety of sources. One source is not enough.
- For Section B you are expected to have undertaken research on your own away from the classroom but under the guidance of your teacher. This should be apparent in your answers.
- Remember to use examples, and to draw diagrams and maps.

Managing questions

Command words

It is important that you understand clearly the meaning of, and difference between, the command words or phrases used in questions. At A2 the style of questioning is more demanding than at AS level and you can expect questions asking for assessment, discussion and examination. Study the list below for the meaning of all the command words.

Assess requires you to weigh up the importance of the topic. There will be a number of possible explanations and you need to give the major ones and then say which you tend to favour.

Evaluate means that, having considered the evidence and looked at the overall explanations for an issue, you must give a point of view. Credit will be given for the justification of the view that you take. It is expected that there will be more than one explanation for any issue.

To what extent and **How far do you agree** both expect explanations for and against and a justification of the view that you favour.

Discuss and **Discuss the assertion** both expect you to build up an argument about an issue and to present evidence for more than one point of view. This type of question expects you to reach a conclusion. Discussion will include both description and explanation and a summary at the end.

Examine asks you to investigate in detail, giving evidence both for and against a point of view or an opinion.

Explain requires you to give reasons or causes and show how, why and where something has occurred.

Compare asks you to point out the similarities, although many versions of such questions expect some **Contrast**. **Contrast** strictly expects the differences. **Compare and contrast** requires both similarities and differences.

Justify is asking for you to state why one opinion or explanation is better than another.

Classify expects you to group ideas or phenomena or explanatory variables into categories.

Use of diagrams and maps

Your answers can be significantly improved by incorporating good-quality maps and diagrams. Work at including appropriate diagrams that summarise key points and provide explanatory annotation that adds support to your text discussion. Maps and diagrams can save words, but they need to be planned as an integral part of your response rather than as an 'add-on' or afterthought at the end of an essay. In an essay that asks for an assessment of coastal management strategies, Figure 1 provides a great deal of information on the topic. It is visually effective and summarises the nature and relative location of a range of management strategies in a specific coastal environment. This would allow you time to develop your assessment of the strategies in more detail in the written part of your response.

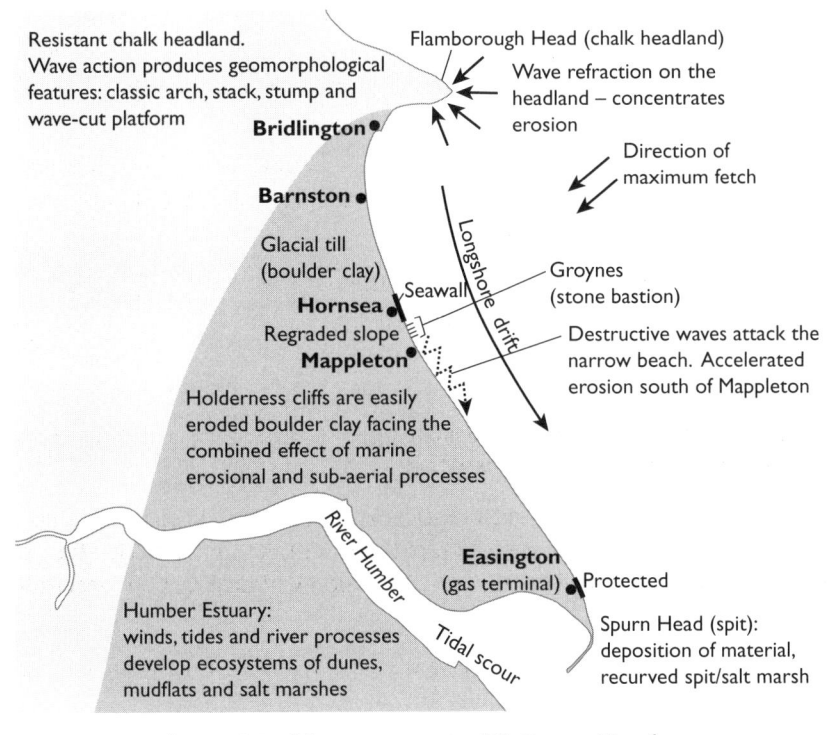

Figure 1 Holderness coast with Spurn Head

Quality of written communication

In addition to assessing your subject knowledge, examiners are required to give credit for the quality of written communication. In this respect examiners are looking for:
- the ability to write clearly and grammatically
- the use of correct spelling
- clear paragraph construction
- appropriate use of geographical terminology
- coherent structuring of the answer with paragraphs linked logically so that the line of discussion being taken is clear to the reader
- reference in the text to supporting diagrams so that they become an integral part of the total response

Revision checklist: Section A

This checklist can be used for any of the themes chosen because for each, six key questions are provided in this book (1.1–1.6 in each case). Put a tick in each box when you are satisfied with your revision for each question and stage.

Contemporary Physical Theme: .... (write your chosen theme from list, p. 12)

Key question	Have organised notes	Have textbook examples	Have own examples	Read texts/ articles/ www	Know terms	Can draw maps	Have diagrams and models	Revision complete
1.1								
1.2								
1.3								
1.4								
1.5								
1.6								

Contemporary Human Theme: .... (write your chosen theme from list, p. 12)

1.1								
1.2								
1.3								
1.4								
1.5								
1.6								

Throughout the Content Guidance that follows, key terms are in **bold**. This should help you to build up your glossary of key terms.

Content Guidance

A2 Geography

Section A is an essay paper that is divided into two parts, each made up of eight questions. Themes 1–3 (Questions 1–8) are set on the following Contemporary Physical Themes:
- Theme 1 Extreme environments: deserts and tundra
- Theme 2(a) Glacial landforms and their management **OR**
- Theme 2(b) Coastal landforms and their management
- Theme 3 Climatic hazards

You will have studied one of the above themes and will be given a choice of two questions on your selected theme.

Themes 4–6 (Questions 9–16) are set on the following Contemporary Human Themes:
- Theme 4 Development
- Theme 5 Globalisation
- Theme 6(a) Emerging Asia: China **OR**
- Theme 6(b) Emerging Asia: India

Again, you will have studied one of the above themes and will be given a choice of two questions on this theme. Each question is worth 25 marks or 40 UMS (total 80 UMS).

Section A Contemporary themes

Theme 1 Extreme environments

1.1 What are the characteristics of a desert environment that make it extreme?

The climatic, biotic and soil characteristics of a desert environment

Hot deserts are located in the subtropical zone between 15° and 35° north and south of the Equator. Examples include the Sahara in north Africa and the Kalahari Desert in the south. The hot desert climate is characterised by high **insolation** rates, subsiding air and high-pressure conditions with outblowing winds. These factors result in exceptionally dry conditions, generally clear skies, and very high evaporation rates, with temperatures during the day averaging over 40°C. On the western land margins, cold ocean currents deprive inblowing marine winds of their moisture before they reach land, contributing to the exceptionally dry conditions and leading to fog conditions offshore, for example, off northern Chile and Namibia. Occasional incursions of the **Inter-Tropical Convergence Zone (ITCZ)** into desert latitudes may allow the penetration of moist winds bringing very heavy, but short-lived, convectional downpours of rain. Average precipitation is 100–200 mm, but effective rainfall is less, primarily due to the very high evaporation rates. **Diurnal** contrasts are more significant than **seasonal** variations as rapid radiation at night, due to clear skies, can result in below-freezing temperatures and heavy morning dew in contrast to the high temperatures and very low humidity conditions during the day.

The productivity of **grey desert soils** is extremely low because the lack of water and high evaporation rates lead to salts accumulating and crystallising on the soil surface. The **organic** layer is very thin as **primary productivity** to produce **litter** is extremely low.

The links between climate, biotic and soil characteristics

Special adaptations are required by biota to overcome the extreme climatic and soil characteristics. The combination of extremes of temperature, potential water loss exceeding annual precipitation and thin, saline soils create challenges for organisms. Adaptation to these extremes includes surviving as an **ephemeral** that takes advantage of short-lived convectional downpours by germinating, growing, flowering and seeding within 20–30 days. Most desert organisms are adapted to withstand or avoid water stress. **Succulents**, such as the cacti of the Americas, have thick cuticles, a low surface-area-to-volume ratio and sunken stomata which open at night to minimise transpiration. These adaptations enable them to survive above ground throughout the year.

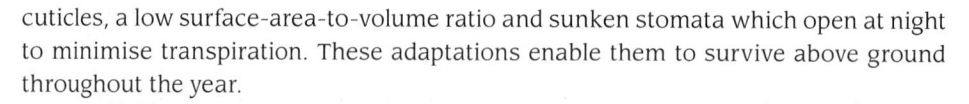

Animals have to adapt to the scarcity of water and food and the temperature extremes. Camels have thick fur to insulate their bodies from the sun, splayed hooves for walking on mobile sands and humps that store fat that provides both energy and water on being respired.

1.2 How is human activity causing pressure on the desert environment?

The threats that are posed by population growth, mineral exploitation, farming and tourism. The positive and negative outcomes of human activity

The desert biome is under threat from a variety of human pressures including population growth, mineral exploitation, farming and tourism. However, the impacts of such activities can be both positive and negative.

Population growth is occurring in desert environments including the Sahara Desert at an average rate of 1.6% per annum. Population growth and greater demand on the land in Tunisia have led to a decline in soil fertility, the lowering of the water table, the silting of dams, and increased flood risk due to new buildings and infrastructure.

The Namib Desert has been sensitive to human activities for decades. The impacts of prospecting and **mineral exploitation** for lithium, beryllium, vanadium and tantalium, as well as more common minerals such as tin, zinc, lead and diamonds have left permanent scars such as the ghost town of Kolmanskop.

In the Sonoran Desert, irrigated **farming** and ranching have been ecologically and economically unsustainable. Vegetation clearing, followed by later field abandonment, the build-up of saline and alkaline soil crusts, pesticide and herbicide use, nitrate fertiliser contamination of stream flows and aquifers, as well as nitrogen enrichment of adjacent wildlands, the introduction and spread of exotic weeds, plant diseases and insects, have all impacted negatively on the desert environment.

On the other hand, agriculture can sometimes have a positive effect on wildlife and fieldside wild plant populations when agriculture is practised on a modest scale, without pesticide and herbicide use.

The increase in **tourism** in the United Arab Emirates, especially involving uncontrolled activity by off-road recreational vehicles (ORVs), and quad and motocross bikes, has had negative impacts on the desert environment and its wildlife. In contrast, the Sinai Peninsula in Egypt offers ecotourists the opportunity to trek with the Bedouin and their camels, causing minimal impact.

1.3 What are the strategies that can be used to manage human activity in deserts?

Strategies that attempt to conserve the desert environment, alleviate the impacts of human activity, control the use of the desert environment and monitor the impacts of human activity

In Tunisia, where 21% of the country is classified as desert and 17.2% is at risk from desertification, a range of strategies have been implemented to manage human activity in the desert environment. Soil and water conservation techniques such as terracing, the construction of earth dams, and slope protection are being implemented by agronomists and hydraulic engineers. Soils degraded through human activity are being treated chemically and enriched with organic matter to counteract their alkalinity. Energy consumption is being altered and rationalised in rural areas to reduce the use of brushwood. In order to monitor the impact of human activity, land degradation maps are being compiled using remote sensing and with the development of a national observatory of land degradation.

The **Sonoran Desert Conservation Plan** is encouraging ranching as an extensive but low-intensity form of land use. Conserving ranches that are currently in operation provides the best way to preserve the integrity of vast tracts of connected, unfragmented open space and wildlife habitat.

The role of local, national and international groups in the management of the desert environment

In Tunisia, various groups are working at different levels to manage the desert environment and avoid it spreading. At the international level, UN institutions such as the FAO (Food and Agriculture Organization) and UNCCD (United Nations Convention to Combat Desertification) have proposed plans that have been adopted at the national level. Tunisia has created a national action plan involving international, national and local organisations. The groups involved at the national level include the Agriculture Ministry, the Ministry for Economic Development and the Environment, and research institutes such as the National Agency for the Protection of the Environment and the National Commission for Sustainable Development. The plan also involves experts and farmers and provides a skills and education action framework. Each region has a development office responsible for environmental protection. The Tunisian national action plan is proving to be effective, but Tunisia has more resources at its disposal than most other African countries.

1.4 What are the characteristics of the arctic and alpine tundra environment that make it extreme?

The climatic, biotic and soil characteristics of a tundra environment

The high latitude of **arctic tundra environments** results in very low temperatures with average temperatures ranging between -5°C and -10°C, long dark winter months when temperatures fall below -20°C and high winds. Tundra regions are dominated by high pressure and subsiding air leading to low mean annual precipitation of below 150 mm. During the short growing season daylight hours are long, but the sun's angle is so low that temperatures rarely rise above 10°C. In summer the ground is permanently frozen apart from the top 50 cm.

The links between climate, biotic and soil characteristics
In arctic areas the growing season is so short that few plants are annuals. Vegetation is dominated by mosses, lichens, grasses and dwarf shrubs. Most plants have short roots so as to avoid the permafrost and small leaves so as to limit transpiration. Dwarf willow and stunted birch trees, with their crowns distorted by the wind, grow adjacent to seasonal rivers but only to a maximum height of 30cm. Plant growth and flowering can be spectacular and colourful during the short growing season. Large numbers of insects appear, giving a few weeks of high productivity. Limited plant growth results in a small amount of litter and the lack of soil biota results in the slow decomposition of organic matter to give a thin layer of peat. When meltwater percolates downwards in the late spring, humic acid releases iron. Underlying permafrost acts as an impermeable layer, causing waterlogging and **gleying**. Bedrock weathered by freeze–thaw action is raised to the surface by frost heave, preventing the formation of soil horizons.

Alpine tundra environments experience low temperatures, high **orographic** precipitation (usually in the form of snow) and high wind speeds. In lower latitudes, alpine tundra environments have a higher diurnal temperature range and frequent freeze–thaw cycles. Soils are usually thin with large amounts of loose rock. Small, brightly coloured alpine plants, such as gentians, grow slowly. Biotic diversity attracts tourists to the Alps at flowering time.

1.5 How is human activity causing pressures on the arctic and alpine tundra environment?

The threats that are posed by mineral exploitation, airborne pollution, global warming and tourism. The positive and negative outcomes of human activity
The tundra biome is under threat from a variety of human pressures including mineral exploitation, airborne pollution, global warming and tourism.

Around Prudhoe Bay in Alaska, the heat produced by buildings and pipelines associated with oil extraction causes permafrost melting. This results in frost heave in the active layer and damage to plants, which take many years to become re-established.

The fragile ecosystems of periglacial environments mean they have a low tourist-carrying capacity. Kuujjuaq in Quebec province, Canada, attracts a small number of tourists interested in hunting and fishing. Nevertheless, irreversible damage has been caused by tourists walking through the tundra.

Tourism in the Alps has grown significantly with the result that tourist and other commercial vehicles are causing dangerous air pollution levels in many alpine valleys. Large areas of permafrost are experiencing greater surface melting associated with the increased atmospheric CO_2 produced by human activity. This creates further positive feedback as bare ground has a lower albedo, which raises surface temperatures and leads to further melting. This melting releases large quantities of methane and leads to further global warming.

You have the opportunity here to show the examiner your ability to make connections between your understanding of human factors and climate change (covered at AS, Unit G1).

1.6 What are the strategies that can be used to manage human activity in arctic and alpine tundra environments?

Strategies that attempt to conserve the tundra environment, alleviate the impacts of human activity, control the use of the tundra environment and monitor the impacts of human activity

The Canadian government has given landmark status to Tuktoyaktuk, an Inuit town in the Northwest Territories that has 1,400 **pingos**, to protect these landforms and the wider area from tourism.

The role of local, national and international groups in the management of the tundra environment

Various groups are working at different levels to manage the tundra environment. At the international level there are global warming initiatives, e.g. Kyoto (1997) and Copenhagen (2009). The Wildlife Management Advisory Council (WMAC) works with the governments of the Canadian province of Yukon and the US state of Alaska, and the Canadian national government, on conserving the wildlife, habitat and traditional use of the Yukon North Slope.

Theme 2(a) Glacial landforms and their management

1.1 What is a glacial system and what are the dynamics of glacial environments?

Glacier mass balance

The inputs to and outputs from a glacier are not constant, but vary over both short and long timescales. The glacier system constantly adjusts to changes in the balance between **accumulation** and **ablation** and this is reflected in the **mass balance** of a glacier. If accumulation exceeds ablation a glacier gains mass (**positive mass balance**). If there is more ablation than accumulation a glacier has a **negative mass balance**.

The impact of climate change on glacier budgets
Glaciers have shown periods of expansion and retreat as climate changes have shifted the net balance to either positive or negative. Here you have the opportunity to show the examiner your ability to make connections between your understanding of glacier mass balance and climate change (covered at AS, Unit G1).

The relationship between climate fluctuations and the geomorphological work done by ice
Glacial landforms can be linked to global events that changed climate. An example of this is the '**Younger Dryas**', a short-lived but significant temperature fluctuation at the end of the last glacial cycle. A large influx of cold fresh water into the North Atlantic resulted in re-growth of glaciers in upland Britain, producing cirque moraines such as Cwm Idwal north Wales.

Cold- and warm-based glaciers, their types and rates of movement
Glaciers can be classified as **cold-based** or **warm-based** depending on whether they are frozen to the underlying bedrock or not. Outside of the polar regions most glaciers are warm-based. However, large glaciers can be cold-based in their upper regions and warm-based near their margins when they extend into temperate climatic zones. Slow rates of **accumulation** and **ablation** associated with glaciers in cold continental climates result in a smaller imbalance between accumulation and ablation and slower ice movement. Glaciers in temperate–maritime climates have greater snowfall in winter and experience more rapid ablation in summer; consequently, the glacier moves more rapidly. There is much more erosion under warm-based glaciers than under cold-based glaciers.

1.2 What are the processes of glacial weathering and erosion and what are the resultant landforms?

Weathering and erosion in the glacial zone
Weathering is the disintegration and decomposition of rocks in situ. The relatively high humidity combined with relatively low temperatures that fluctuate above and below freezing make **freeze–thaw** weathering, a **physical weathering** process, predominant. The low temperatures make **chemical weathering** less important. Generally freeze–thaw weathering contributes to the formation of angular features in glacial environments, such as **arêtes**. Freeze–thaw weathering on slopes produces rock slides leading to **screes**, but higher temperatures in summer may lead to mudflows and soil creep, known as **solifluction**. Freeze–thaw weathering and mass movement of scree contribute directly to the formation of various types of **moraines** and provide glaciers with the abrasive material that makes these agents effective. Weathering is very important as it allows the ice to move material that has already been loosened.

The processes of **glacial erosion** can be divided into three categories: glacial **plucking**, glacial **abrasion** and glacial **meltwater** erosion.

1. **Plucking** occurs when ice freezes around protruding rocks that are then plucked away as the ice moves. This process is particularly effective on well jointed rocks and in previously weathered areas, such as the backwalls of cirques.

2. **Abrasion** involves the 'sandpapering' effect of angular material embedded in a glacier's sides and base. This is only effective when there is a continuous supply of moraine.

3. Glacial **meltwater** erosion is not, strictly speaking, a glacial process, but removes material during spring and summer melting.

Landforms of glacial erosion to include macro-scale, meso-scale and micro-scale landforms

The characteristic macro-scale landforms produced by erosional processes include **cirques**, **arêtes**, **pyramidal peaks**, **glacial troughs**, **hanging valleys**, **truncated spurs** and **crag-and-tail** landforms.

Meso-scale landforms include **roches moutonnées** and **subglacial meltwater channels**.

Micro-scale landforms include **striations**.

Cirques are basin-shaped mountain hollows formed by small cirque glaciers that are characterised by rotational movement. Their steep head- and sidewalls result from freeze–thaw weathering and plucking, and their smooth and gentle floors are the result of abrasion. Under pressure from the volume of ice, cirque glaciers rotate sufficiently to move uphill and over the **rock lip** at the exit from the cirque, e.g. as at Cwm Cau on Cadair Idris mountain (Figure 2).

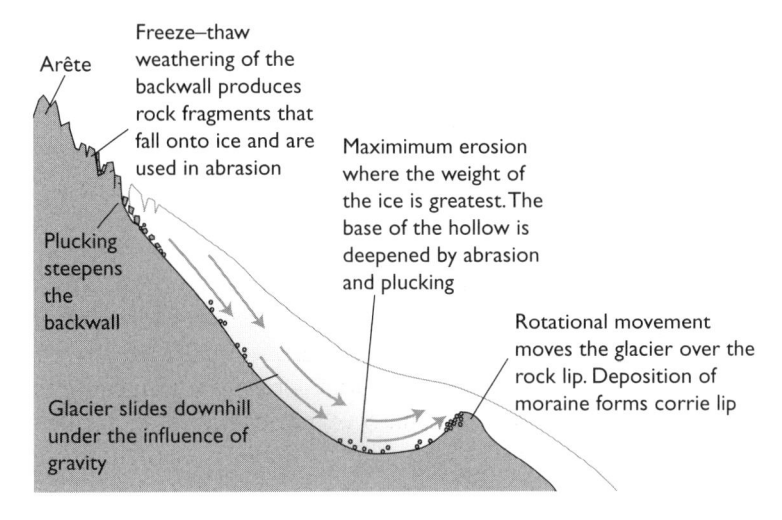

Figure 2 Cross-section to show glacial processes responsible for cirque formation

Arêtes are narrow, knife-edged ridges resulting from the headward erosion of neighbouring cirques through freeze–thaw weathering and plucking, e.g. Crib Goch, Snowdonia.

Pyramidal peaks form as a result of headward erosion by three or more cirque glaciers, e.g. Mount Snowdon.

Glacial troughs start with a pre-glacial V-shaped river valley which is deepened and has its sides steepened by the processes of freeze–thaw weathering, abrasion and plucking, to eventually form a U-shaped glacial valley, e.g. Nant Ffrancon valley, Snowdonia.

Hanging valleys relate to pre-glacial tributary streams feeding the main river channel. Glacial overdeepening of this main valley leaves the less eroded tributary valley hanging over the main trough, with its river as a waterfall cascading over the edge.

Truncated spurs indicate how powerful glacial erosion is compared to fluvial erosion, as the interlocking spurs of pre-glacial river valleys are eroded by powerful valley glaciers that follow an essentially straight course.

Crag-and-tail landforms (Figure 3) develop where a glacier overrides a mass of hard rock (the crag), which protects softer rocks in its lee. The latter form a tapered, gently sloping ridge (the tail). This is a good example of **differential erosion**.

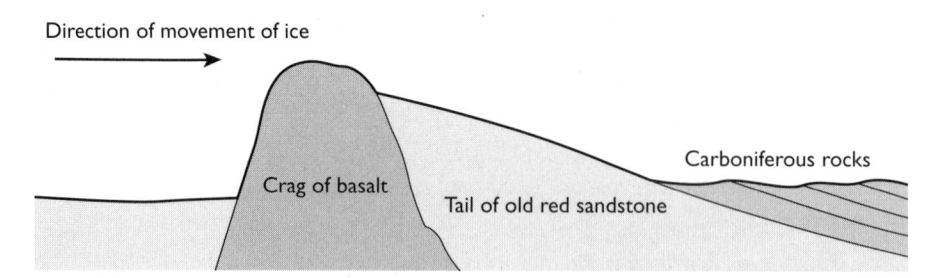

Figure 3 Crag-and-tail landforms

Roches moutonnées develop when a large block of resistant rock is eroded. The upglacier (**stoss**) slope shows evidence of abrasion in the form of striations, and slopes gently. The downglacier (lee) face is steeper and rougher due to plucking (Figure 4).

Subglacial meltwater channels develop where meltwater is channelled beneath a glacier, widening and deepening existing grooves, e.g. Cheriton Valley, Gower.

Striations are abrasion marks on rock surfaces that indicate the direction of ice flow.

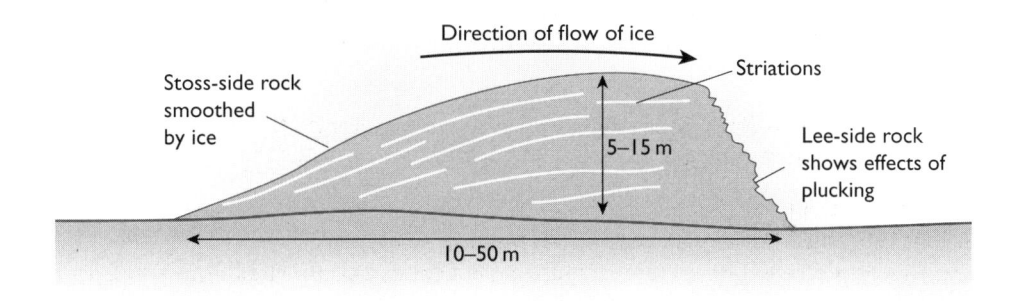

Figure 4 A roche moutonnée (section)

1.3 What are the processes of glacial transport and deposition and what are the resultant landforms?

Transport and deposition in the glacial zone

The transportation of load by glaciers is dependent on the depth of the ice, the amount of load carried and the temperature and pressure conditions that affect the 'fluidity' of the ice slope gradient. Transported material (debris) can be classified as either **supraglacial** (on the glacier surface), **englacial** (within the glacier), or **subglacial** (beneath the glacier). Subglacial debris is the most altered during transport. Deposition is dependent on changes in levels of energy in relation to the load carried. This may be affected by changes in atmospheric processes, so in periods of higher temperature and less snowfall ice movement is reduced, leading to lower energy levels and more deposition. The processes by which glaciers deposit material are complex.

Landforms of glacial and fluvioglacial deposition

The characteristic landforms produced by depositional processes include subglacially formed moraines such as **drumlins** and **till plain**, and ice-marginal moraines such as **terminal**, **recessional**, **lateral** and **medial** moraines. In periods of higher temperature when ice and snow melt occurs, fluvioglacial deposition is important and creates **eskers**, **kames** and **outwash gravels**. **Kettle holes** or **kettle lakes** may also occur when the ice beneath melts.

Drumlins are smooth, oval hills that occur in swarms, creating 'basket-of-eggs topography'. They are likely to have formed under deep, mobile ice and their degree of elongation is related to the speed of ice movement (Figure 5). In Ireland a belt consisting of tens of thousands of tightly-packed drumlins extends from County Down in Northern Ireland, through County Mayo to Donegal Bay in the Republic of Ireland.

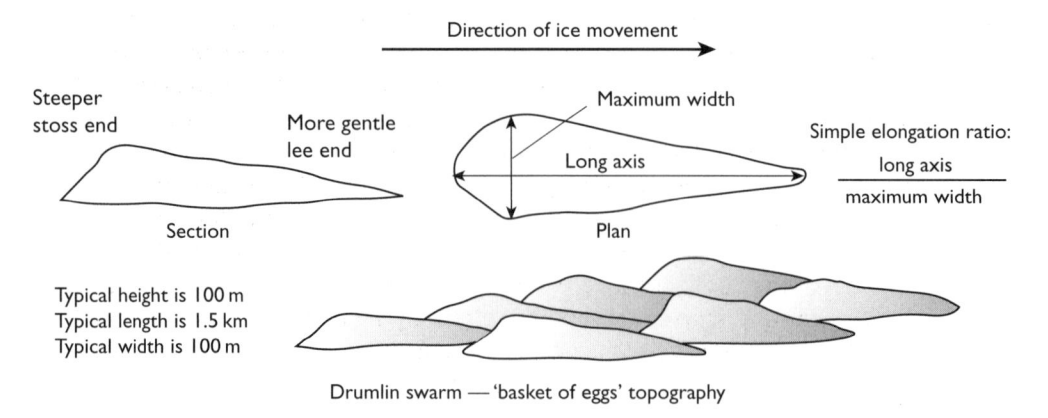

Figure 5 Drumlins, showing stoss and lee ends and drumlins' orientation

Till plain is a combination of subglacial, englacial and supraglacial moraine released by a stagnant glacier as it melts, which leaves the debris unsorted.

Terminal moraines mark the maximum advance of the glacier or ice sheet, e.g. the Glais moraine in the Swansea Valley.

Recessional moraines form where the ice front remained stationary for a period before retreating further.

Lateral moraines are embankments of freeze–thaw weathered debris running along the sides of a glacier's valley. **Medial moraines** result from the merging of lateral moraines from two tributary glaciers joining together.

Eskers are winding ridges of silt, sand and gravel laid down by meltwater in a sub-glacial tunnel orientated approximately at right angles to the ice front.

Kames are irregular mounds of stratified sands and gravels, formed from a decaying glacier or ice sheet.

Outwash gravels are horizontally bedded sheets of gravel deposited by summer melt or deglaciation downward of the ice front.

Kettle holes are circular depressions, initially filled by meltwater, resulting from the gradual decay of a block of ice buried by overlying sediments.

1.4 What are the effects of deglaciation on the landscape?

The effects of deglaciation on the landscape to include periglacial, fluvial and sub-aerial processes

Deglaciation is the reduction in size of glaciers and ice sheets resulting from a negative mass balance. It leads to the exposure of a previously ice-covered surface. It includes the retreat of periglacial processes and landforms to higher altitudes and latitudes. In areas of low relief, important periglacial processes are **frost heaving** and **thrusting**. Associated periglacial landforms are **pingos** and **patterned ground**. On slopes, important periglacial process are **freeze–thaw weathering** and **solifluction** and associated periglacial landforms are **blockfields**, **scree slopes** and **solifluction lobes** and **benches**. Relevant geomorphological processes include **mass movement** processes (modifying valley profiles largely created by glacial erosion), fluvial processes (resulting in the infilling at the head of ribbon lakes), or weathering processes (breaking down glacial and fluvioglacial deposits). Since the last glaciation the change to temperate conditions, together with changes in **base level** due to **isostatic** adjustment, have significantly modified glacial landforms. Other examples of the influence of isostatic adjustment are raised erosion surfaces in glaciated uplands and rejuvenation features within glaciated valleys.

Again, show the examiner your ability to make connections between your understanding of deglaciation and climate change (covered at AS, Unit G1).

1.5 Why are glacial environments important?

The impact of glacial processes and landforms on human activity

Glacial processes impact on human activity because of the high incidence of **avalanches**, **rock falls** and other forms of mass movement such as **landslides** and **glacial outburst floods**. On 23 February 1999, 31 people were killed by an avalanche in Galtür, Austria, which took only 50 seconds to reach the village.

Glacial landforms (in areas that are currently experiencing glaciation and in formerly glaciated areas) present constraints and provide opportunities for human activity in terms of tourism, water supplies and energy, agriculture, mining and quarrying, settlement and corridors for transport.

The impact of human activities on glacial environments

Some of the impacts of human activities on glacial environments are:
- Leisure activities — winter-sports activities, including associated infrastructure such as buildings, ski lifts and road access.
- Logging activities leading to the removal of vegetation cover, which accelerates weathering and mass movement processes.
- Damming of glacial lakes for use as reservoirs for hydro-electric power schemes.
- Pollution and permafrost degradation through settlement and heat and waste disposal.
- Anthropogenic climate change, leading to the net ablation of glaciers worldwide.

Opportunities and limitations for human activity presented by the shift of the permafrost limit

Opportunities for human activity presented by the shift of the permafrost limit include settlement and the development of mining and oil extraction industries. Limitations for human activity include damage to structures caused by freeze–thaw in the active layer and ground subsidence.

1.6 What are the methods used to manage glacial environments and how successful are these strategies?

Management of the impacts of glacial processes and landforms on human activity

Methods used to manage the impacts of glacial processes on human activity include prevention or control measures in the form of **soft** and/or **hard engineering** strategies. These include constructing strong, resistant buildings; constructing **avalanche barriers** on mountain slopes; and **planting trees** to break the flow of an avalanche. In response to the loss of life caused by the 1999 Galtür avalanche, a barrier wall was constructed. The choice of strategy is dependent on the nature of the human activity; the density of human settlement; the nature of the impact; the frequency of occurrence and degree of intensity of impact, loss of life and injury; and the damage caused to property and infrastructure.

Management of the impacts of human activities on glacial environments

Strategies used to manage the impacts of human activities on glacial environments include **prevention of access** to and **bans on use** of (aspects of) the area under impact. The occupation of land and character of land use in glacial environments can be controlled by means of **planning controls** and **zoning**, which affect access, location and design of buildings, and infrastructure. In Snowdonia National Park a variety of strategies, including repairing eroded upland footpaths, has been implemented to manage the impact of human activities in a formerly glacial environment.

Assessment of the success of strategies for managing either glacial processes/landforms or human activities

The assessment may include a financial cost–benefit analysis. This examines the extent to which strategy aims are achieved in terms of controlling or lessening undesired impact(s); the projected life of the management strategy or strategies; or the extent to which each involved agency or body feels that any implemented strategies have achieved objectives. Finally, having considered the interests of the different participants and the actual evidence 'on the ground', a summary evaluation of the success or lack of success of the strategy or strategies needs to be made.

Theme 2(b) Coastal landforms and their management

1.1 What is a coastal system and what are the dynamics of coastal environments?

The coastal system

The coastal system is one of **inputs** and **outputs**. There are two systems:

1. The **cliff system,** with *inputs* comprising the sub-aerial processes of **weathering** and the atmospheric process of wind erosion; a *throughput* comprising cliff **mass movement** of **falls**, **slips** and **slumps**, and an *output* of sediment at the base of the cliff, which is either deposited or transported by marine processes.

2. The **beach system**, with an *input* of sediment from longshore drift, the cliff and offshore, a *throughput* of longshore drift and an *output* of longshore drift and destructive waves carrying sediment offshore.

Coastal sediment cells

Coastal **sediment cells** are areas of coast usually defined by headlands within which marine processes are largely confined, with limited transfer of sediment from one cell to another. They vary in size depending on the nature of the coast.

The state of dynamic equilibrium in the coastal system

The relationship between inputs and outputs is constantly changing — it is dynamic — and the system works towards an equilibrium position where inputs equal outputs. Erosion, transport and deposition occur, giving rise to the concept of **dynamic equilibrium**.

Wave types and characteristics and their variations over time and space

There are two extreme forms of wave: **destructive** and **constructive** waves. They have different characteristics and occur in different places determined by the local configuration of the coastline and/or prevailing wind conditions.

1. **Constructive waves** are low, flat and gentle, with wavelengths up to 100 m and a low frequency of 6–8 waves per minute. They are characterised by a relatively more powerful **swash**, which carries sand and shingle up the beach, and a relatively weaker **backwash**. Constructive waves contribute to the formation of beach ridges and **berms** (Figure 6).

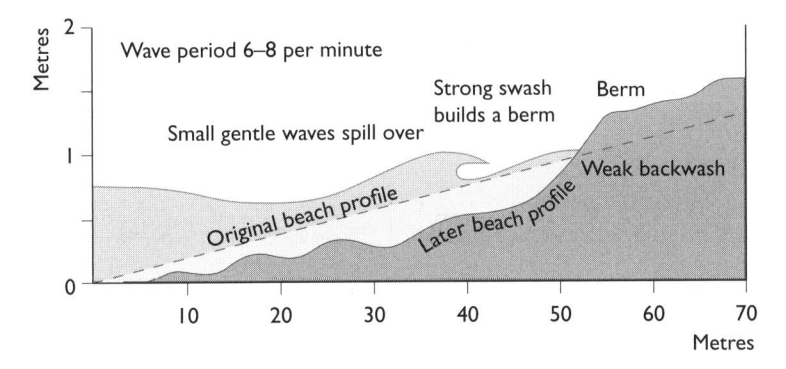

Figure 6 Constructive waves

2. **Destructive waves** tend to occur during storms, are steep in form and break at a high frequency, at 13–15 waves per minute. They have a plunging motion that

generates little swash and a relatively more powerful backwash; this transports sediment down the beach face, resulting in a net loss of material (Figure 7).

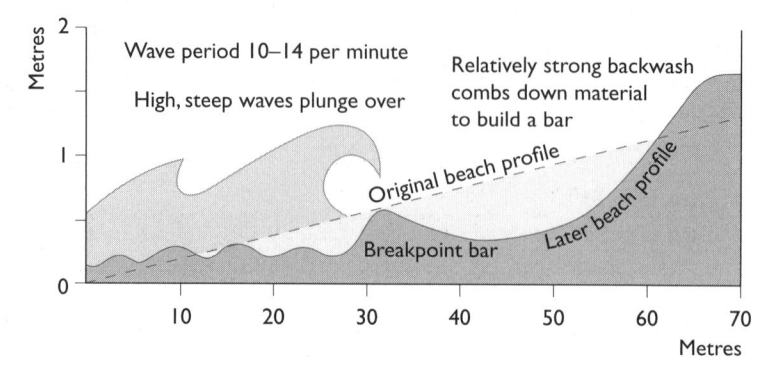

Figure 7 Destructive waves

Most beaches experience the alternating action of constructive waves in summer and destructive waves in winter, resulting in an annual cycle of beach growth and decay.

Wave refraction concentrates wave energy on headlands, which causes erosion, while dissipating energy in bays, which causes deposition.

1.2 What are the processes of coastal erosion and what are the resultant landforms?

Weathering and erosion in the coastal zone

Weathering in coastal environments includes physical disintegration caused by such processes as **freeze–thaw**, **salt crystallisation** and **wetting and drying**; chemical decomposition includes **solution** and **carbonation**. The variety of intertidal organic life encourages **biotic** weathering, with activity ranging from that of the roots of seaweed to acid secretions by limpets, barnacles and seagulls. Processes of erosion in coastal environments include **corrosion**, **hydraulic action**, **abrasion** and **attrition**.

Landforms of coastal erosion

Cliffs are rocky faces that develop along coastlines where marine undercutting has been ongoing. Undercutting at the base of cliffs leads to the formation of a **wave-cut notch**, then collapse by fall or slumping, depending on the **lithology** and **dip** of the rocks (see 1.4 below). Progressive retreat of the cliff will leave a gently sloping intertidal **wave-cut platform**.

Caves develop where lines of weakness such as joints or faults are exploited. **Arches**, e.g. Durdle Door in Dorset, form on headlands where caves erode back-to-back. **Stacks**, e.g. Old Harry, also in Dorset, result when the arch collapses. **Stumps**, e.g. Harry's Wife — located next to Old Harry — remain after the stack collapses (Figure 8).

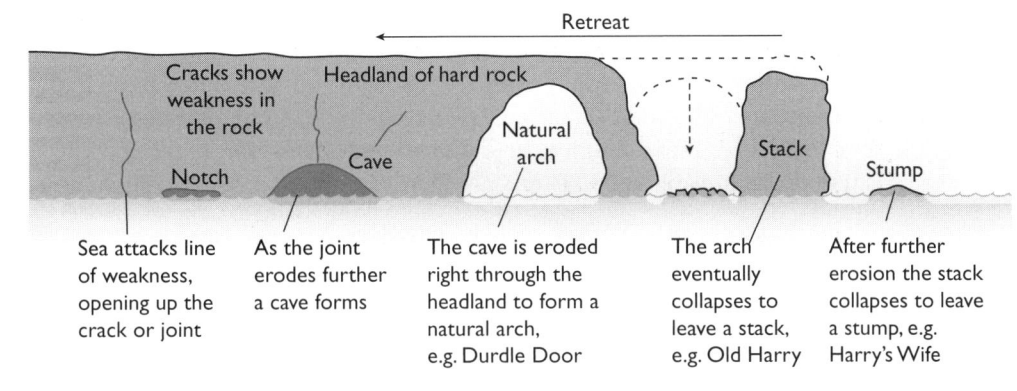

Figure 8 Cave–arch–stack–stump erosional sequence

Sea attacks line of weakness, opening up the crack or joint

As the joint erodes further a cave forms

The cave is eroded right through the headland to form a natural arch, e.g. Durdle Door

The arch eventually collapses to leave a stack, e.g. Old Harry

After further erosion the stack collapses to leave a stump, e.g. Harry's Wife

Sea-level rises and erosion

Landforms created with rises of sea level are associated with either **eustatic** rise after a glacial retreat, as with the **Flandrian transgression**, or local areas of **subsidence**, as in the case of the south-east of England. Landform features include those associated with marine erosion and the submergence of the land by an encroaching sea, such as **rias** and **fiords**. For example, the **raised beaches** of Gower are wave-cut platforms eroded during a period with higher sea levels.

1.3 What are the processes of marine transport and deposition and what are the resultant landforms?

Transport and deposition in the coastal zone

Beach sediment is transported progressively along a beach by the process of **longshore drift**. This is caused by the oblique approach of waves and consequently swash, followed by the direct return to the sea of the backwash (Figure 9).

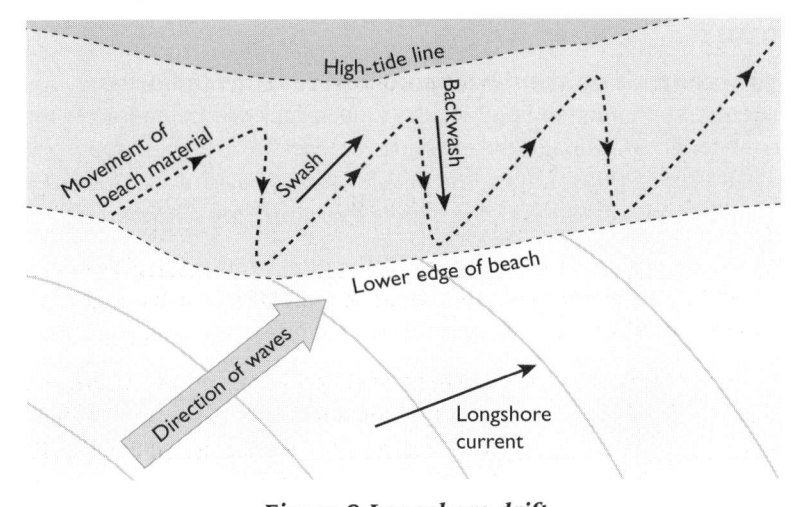

Figure 9 Longshore drift

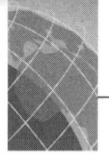

Landforms of coastal deposition

Spits are banks of sand and shingle projecting from the shoreline into the sea. They need a supply of sediment from longshore drift to build and maintain them. Often the far end is **hooked**, being formed either by wave refraction or local wave approach from a different direction. **Double spits** occur where longshore drift extends one spit in the direction of the prevailing winds and another from the opposite direction, as at Poole in Dorset.

A **tombolo** is a spit joining an island to the mainland. For example, Chesil Beach links the Isle of Portland to the Dorset mainland.

Spits may grow across small bays, eventually closing them in with complete sand ridges called **barrier beaches**, e.g. Slapton Sands in Devon.

Bay-head beaches are areas of sand or shingle beach occupying part of a bay bounded by projecting headlands, e.g. Barafundle Bay in Pembrokeshire.

Offshore bars are ridges of sand and/or shingle developed offshore on a gently shelving coastline.

Cuspate forelands are triangular outgrowths of shingle ridges formed by longshore drift from opposite directions, e.g. Dungeness in Kent.

Sea-level rises and deposition

Sea-level rises will create **coastlines of submergence**. Eustatic rises in sea level are associated with glacial retreats and give rise to inundation of low-lying areas by the sea. **Estuaries** are created, which at low tide expose mudflats and deposited material.

Do you have your own examples of all these landforms?

1.4 What is the role of geology in the development of coastal landforms?

Lithological controls on the development of coastal landforms

Beach material is often made up of locally eroded rock whose character influences beach characteristics. Shingle, for example, can retain a higher slope angle than sand and allow greater infiltration. The headlands and bays on the Gower Peninsula are produced by alternate layers of resistant **carboniferous limestone** rocks and weaker **'Namurian' shales** (Figure 10).

Igneous rocks such as **granite** erode more slowly and produce steep-sided cliffs, e.g. Land's End. Cliffs composed of **unconsolidated sands** and **clays**, e.g. at Barton-on-Sea, Hampshire, suffer from the complex mass-movement processes of slides, slumps and flows.

Before erosion

After erosion

- Limestone
- Anticline
- Shale
- Syncline

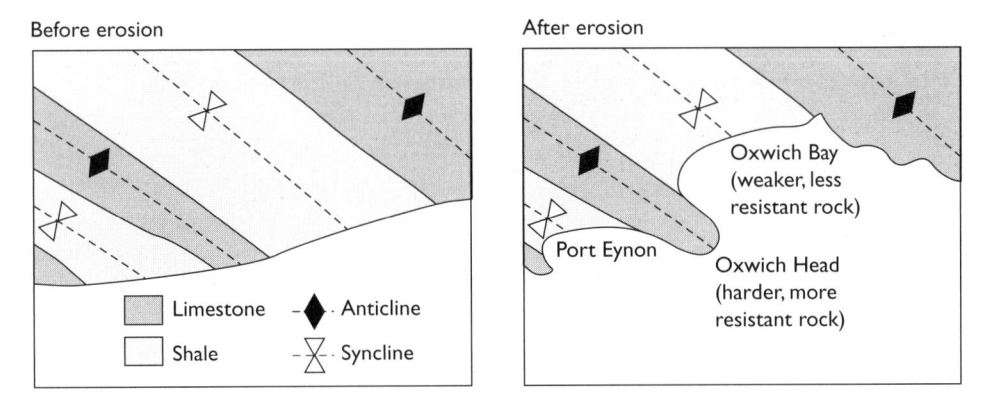

Oxwich Bay (weaker, less resistant rock)
Port Eynon
Oxwich Head (harder, more resistant rock)

Figure 10 Headlands and bays (Gower Peninsula)

Structural controls on the development of landforms

Geological **structures** incorporating bedding planes, faults and cracks can add distinctive features to coastal cliff lines, such as the shape of caves and other local features like **blowholes** and **geos**.

The orientation of the coastline with the local geology is a very important factor (Figure 11). If the geological trend is **concordant**, i.e. parallel to the coast, then a **Dalmatian coastline** of coves and solid rock bars is created. Differential geology at right angles to the coast will result in a **discordant coastline** with bays and headlands, e.g. Isle of Purbeck, Dorset.

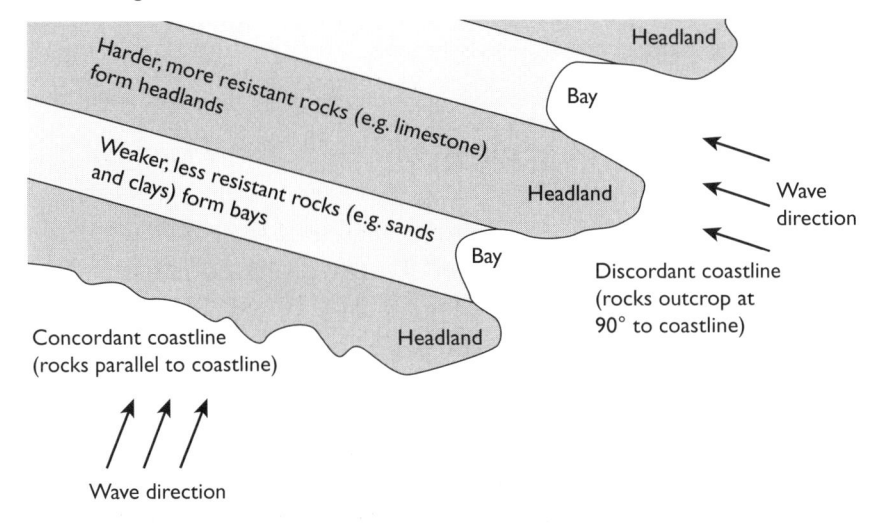

Harder, more resistant rocks (e.g. limestone) form headlands
Weaker, less resistant rocks (e.g. sands and clays) form bays
Headland
Bay
Headland
Bay
Headland
Wave direction
Discordant coastline (rocks outcrop at 90° to coastline)
Concordant coastline (rocks parallel to coastline)
Wave direction

Figure 11 Concordant and discordant coastlines

1.5 Why do coastal environments need to be managed?

The impact of coastal processes and landforms on human activity

Coastal processes can impact on human activity in both a negative and positive way. Processes of weathering and mass movement along the coastline can endanger buildings, as with the collapse of the Holbeck Hall Hotel in North Yorkshire in 1993. Transport of material can silt up estuaries and make harbour entrances more difficult to navigate, e.g. Poole Harbour in Dorset, and the Newhaven ferry terminal on the River Ouse estuary in East Sussex. However, deposition also creates beaches that can be used for tourism, e.g. Studland Bay, Dorset.

Coastal landforms impact on human activity. Coastal lowlands are important for food production. Coastal ports play a leading role in world trade, and the coast has become a popular place for leisure, recreation and tourism.

The impact of human activities on coastal environments

Important large-scale industries that use a lot of space often have coastal locations, e.g. steelworks at Port Talbot, oil refineries at Milford Haven, gas terminals at Easington, Yorkshire, and defence installations at Plymouth.

Seawalls, jetties and docks will be created to serve these installations and these will alter local tidal currents and influence local erosion and deposition patterns.

Tourism will have an impact on processes in cases where local authorities install seawalls, promenades and groynes to build sandy tourist beaches. In the Holbeck Hall Hotel case, for example, the building itself could have played an instrumental role in its own collapse because the extra weight of the structure could have made the underlying clay more susceptible to slumping.

Negative impacts of leisure and recreation activities on coastal environments include footpath erosion and people trampling on areas with fragile coastal ecosystems, such as dunes. Offshore dredging of sands and gravels may also affect the supply of materials to beaches.

1.6 What are the methods used to manage coastal environments and how successful are these strategies?

Management of the impacts of coastal processes and landforms on human activity

Hard engineering strategies include:
- Concrete **seawalls** that reduce wave energy but are expensive to build and maintain.
- **Rock armour (rip-rap)**, consisting of large boulders of hard rock placed in front of seawalls and sometimes used as groynes. These are both expensive and ugly.
- **Groynes**, which are walls built at right-angles to the coast. Traditionally they have been made from wooden railway sleepers, but they can be made of rock armour. They prevent beach migration.

- **Gabions**, which are blocks made by wire-netting together medium-sized pieces of hard rock. These are expensive and can be ugly.
- **Revetments**, which are slatted and angled low wooden walls built parallel to the beach. They act to absorb wave energy and protect soft cliffs. They are ugly and liable to rapid damage.

Cliff regrading does limit slumping, particularly when combined with cliff **drainage schemes**, but this does not stop the loss of beach material.

Soft engineering strategies include:
- **Beach nourishment**, which is the artificial input of beach material to compensate for natural losses. Miami Beach in Florida is managed in this way, but such a strategy is extremely expensive.

The Holderness coast, in the East Riding of Yorkshire, is in active retreat and coastal management strategies are critical. The village of Mappleton is threatened by coastal erosion and its access road is within 50 m of the cliff edge. In 1991, **hard engineering** was undertaken to slow down the rate of erosion. A **groyne** made from granite boulders was built out to sea, behind which sand was trapped to form a beach. Another groyne and a **revetment** were constructed south of the first groyne and at right-angles to it to absorb the destructive energy of waves and restrict the removal of beach material by longshore drift. There has also been an attempt to stabilise the boulder-clay cliffs by **regrading** the cliff face so as to reduce its angle and thereby reduce mass movement (Figure 12).

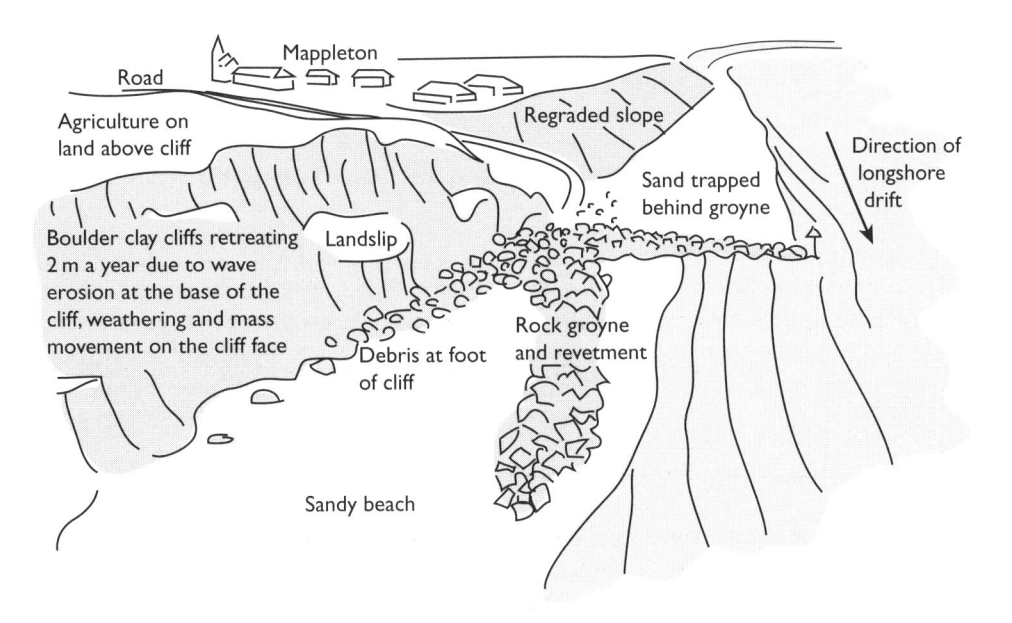

Figure 12 Coastal management strategies at Mappleton, Holderness

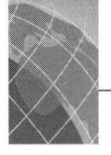

Management of the impacts of human activities on coastal environments

Lulworth Cove in Dorset receives around 750,000 visitors every year. However, there are many problems associated with tourism, including large car parks needed for vehicles, unsuitable or unsightly tourist shops, footpath erosion, noise and air pollution, rubbish and sewage, and erosion of the Geological SSSI associated with fossil-hunting and field trips.

Management of these human activities includes footpath management; information for tourists, such as display boards, leaflets, guided walks and talks and educational displays; the screening-off of 'eyesore' sites; and the setting-up of a coastal volunteer system. Have you got an alternative example?

Assessment of the success of strategies for managing either coastal processes/landforms or human activities

The management strategies put in place at Mappleton have reduced coastal erosion but not solved the problem. Local people are demanding more protection but the government and local authority are reluctant to undertake more work as they believe the costs of further coastal management outweigh the benefits. Additional work would only slow down, not stop, further coastal retreat. Also, trapping more sediment at Mappleton would create more erosion further south, because less sediment would be transported by longshore drift to build up beaches and protect the coast.

Theme 3 Climatic hazards

1.1 How does global atmospheric circulation give rise to global climatic zones?

Atmospheric movement

Solar energy (**insolation**) powers the atmospheric system and the energy circulations within it. The amount of solar energy (**heat budget**) received varies with latitude. The tropics have an **energy surplus** as they gain more from insolation than is lost by **radiation**. The higher temperate and polar latitudes have an **energy deficiency**, losing more by radiation than is gained by insolation. This imbalance in energy distribution sets up a transfer of heat energy from the tropics to higher latitudes.

The tricellular model: the Hadley, Ferrel and polar cells

This global transfer of energy is the basis of global atmospheric circulations, which give rise to the low- and high-pressure belts and the planetary wind systems associated with the Earth's three major convection cells: the **Hadley**, **Ferrel** and **polar cells**. These make up the **tricellular** model that controls atmospheric movements and the redistribution of heat energy (Figure 13).

WJEC Unit 3

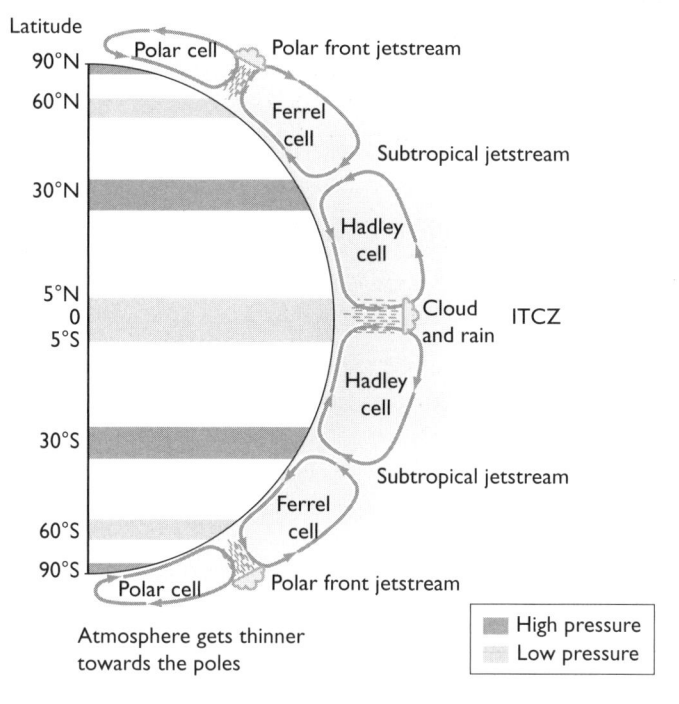

Figure 13 Convection cells and pressure belts

The patterns of winds and the world's pressure belts

This is illustrated by Figure 14.

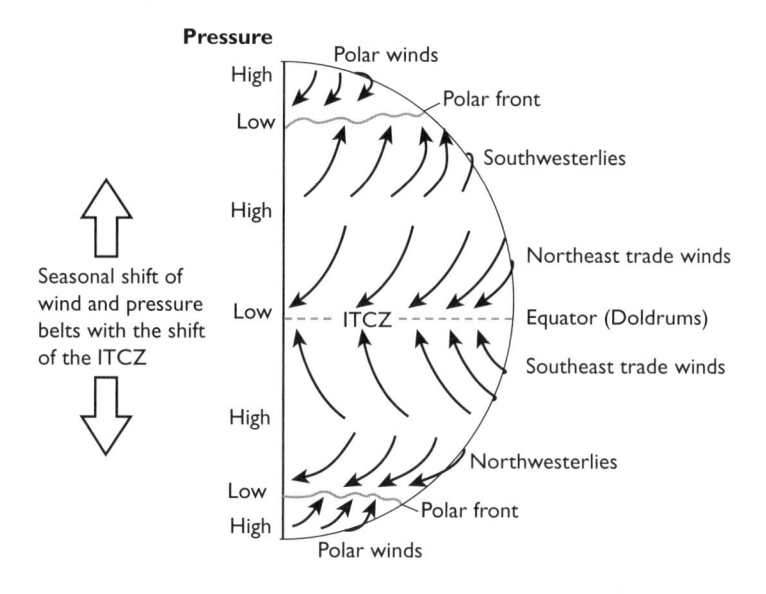

Figure 14 Pressure belts and associated wind systems

33

1.2 Why do seasonal and periodic variations of climate occur?

Seasonal variations

The reasons for seasonal variations in climate are:
- The seasonal movement of the **Inter-Tropical Convergence Zone (ITCZ)**, and pressure and wind belts associated with the movement of the sun's overhead position during the year.
- The effects of the warm and cool ocean currents.
- Temperature differences between continental landmasses and neighbouring ocean waters.

In the examination, you only need to refer to one climatic type from either a **tropical** or **temperate** region. In the tropical region seasonal changes are far more marked in **savanna** and **monsoon** climates. In the temperate region seasonal changes are more pronounced for the **continental interior** and **east-coast margin**.

The following is an explanation of seasonal variations in **monsoon** climates.

Monsoon climatic type

Such climates occur mainly on the eastern side of continental landmasses in the tropics, extending across approximately 5–20 degrees of latitude.

The type is marked by a *distinct hot wet and a cooler dry season*, determined by the annual movement of the ITCZ between the tropics and associated movement of pressure belts and seasonal reversal of winds consequent on this. The monsoon climate regime is most clearly seen in the Indian subcontinent, but exists in other regions north and south of the Equator on the eastern edge of continents, e.g. east Africa.

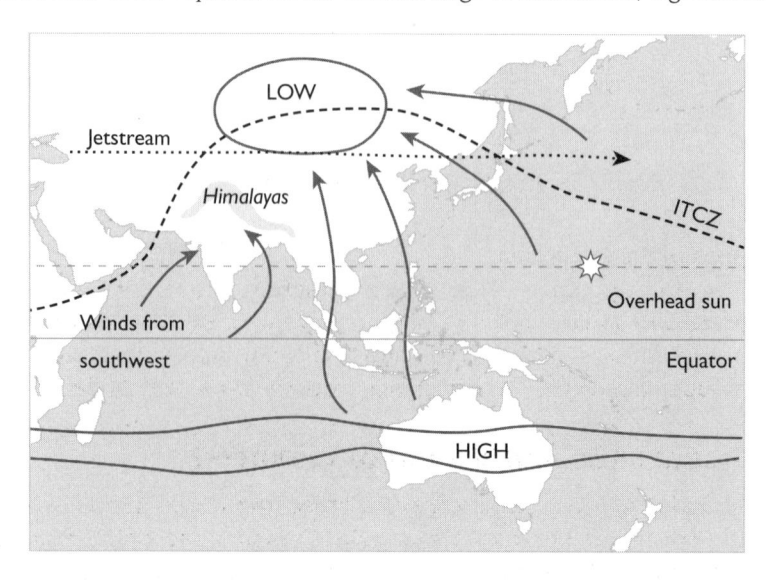

Figure 15 The wet monsoon season (June to October)

The *wet monsoon season* (Figure 15) occurs with the movement of the ITCZ into the region. This brings an area of low pressure and draws in hot, moist winds from the ocean. Rainfall is increased by **orographic uplift**, where these moist winds are drawn over uplands, e.g. the Western Ghats in India. Temperatures are high, averaging 30°C. Humidity is also very high, with average rainfall around 2000mm, decreasing with distance inland. Cyclones and hurricanes are frequent towards the end of the rainy season.

The *cooler dry season* (Figure 16) coincides with the extension of continental high pressure as the ITCZ moves back towards the Equator and across into the tropics beyond. With high pressure dominating, there is air subsidence and the outblowing winds are dry. Temperatures remain relatively high, at 25°C in lowland areas, and evaporation rates are also high. The weather is much more severe in mountain areas.

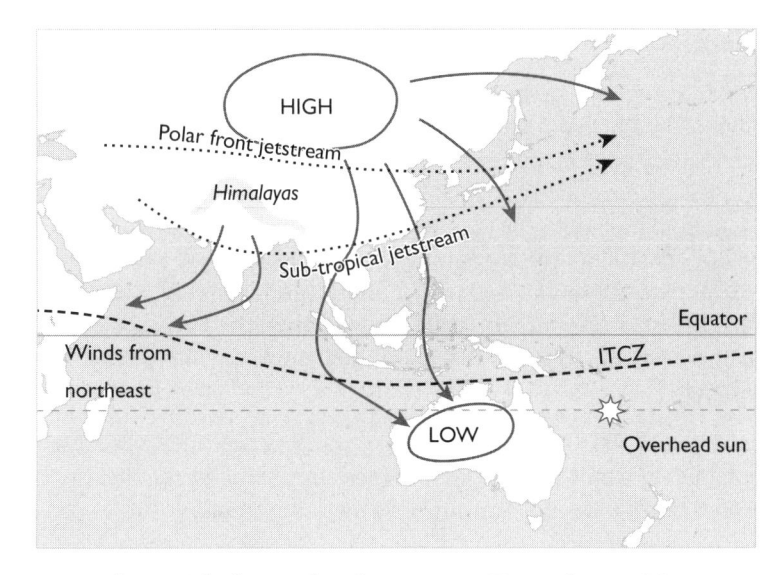

Figure 16 The cooler dry season (November to May)

Periodic changes in climate
Periodic variations in climate occur over both the long term and short term. **Glacials** and **interglacials** are examples of long-term changes. **El Niño/La Niña** cycles are examples of short-term changes. You have the opportunity here to show the examiner your understanding of climate change (covered at AS, Unit G1).

1.3 What are the world's major climates?

For the examination, you need to have a broad knowledge and understanding of the world's major climates, including the distribution of the main climate types for the tropical and temperate latitudinal belts. However, detailed reference needs to be made to only ONE climatic type chosen from either a **tropical** or **temperate** region.

The main climatic types in tropical regions
The main influences on the climates of **tropical** regions are:
- The overhead or near-overhead position of the sun giving high insolation throughout the year.
- The position and seasonal movement of the ITCZ together with the tropical pressure belts' wind systems.
- The path of the upper jetstreams affecting the path of low-pressure systems.
- Differential heating of landmasses and oceans in the tropics affecting air-pressure patterns and seasonal wind directions.
- The effects of offshore cold currents on western land margins and warm currents on eastern margins.
- The position of mountain ranges and their effects on incoming moist winds off the ocean.

Savanna climatic type
5–20 degrees latitude either side of the Equatorial belt.

High temperatures of 35–25°C prevail all year because insolation is high. This climate type is distinguished by having a *hot, wet* season and a *cooler, dry* season. Humidity is highest in the wet season, but evaporation rates remain high during the cooler dry season. Rainfall occurrence is associated with the movement of the ITCZ towards the tropic in association with the apparent movement of the sun's position overhead. During the hot season, as this occurs, low pressure prevails with moist inblowing winds and rising air currents leading to **convection** rainfall. Rainfall amounts are most reliable towards the Equatorial latitudes, where they average 800 mm a year, but become less reliable towards the hot desert margins, where they average 300–400 mm annually.

The cooler, dry season in the savanna belt occurs at the time when high pressure and dry, outblowing winds prevail; this is when the overhead sun and the ITCZ move away to extend beyond the Equator towards the other tropic.

The main climatic types in temperate regions
The main influences on climate in **temperate** regions are:
- Mid-latitude position.
- The influence of the mid-latitude low-pressure belt and the atmospheric conditions along the polar front as well as the influence of the upper jetstream, except for continental areas in winter.
- The seasonal shift of the pressure and wind belts.
- The position and interaction at the margins of the different air masses affecting areas in temperate latitudes.
- Differential heating of the continental interior and ocean margins.
- The effects of ocean currents and the air above them.
- The location of upland ranges in relation to prevailing winds.

Maritime west margin European climatic type
Mid latitude, 35 to 55 degrees.

This climatic type is characterised by relatively mild temperatures (average seasonal range 5–20°C), along with high humidity and precipitation (averaging 600mm) throughout the year. However, precipitation totals are significantly higher over upland areas in the face of prevailing moist westerly winds coming off the ocean, e.g. in the Cambrian Mountains of Wales. Conversely, precipitation totals are low in rain-shadow areas, e.g. lowland East Anglia.

The temperatures and precipitation figures are mainly influenced by the mid-latitude position, low-pressure belt and the mild westerly prevailing winds. The latter are warmed by warm currents, e.g. the Gulf Stream, on the west margin of landmasses.

The weather is strongly influenced by the position of the polar front, the associated jetstream, and the passage of westerly-moving depressions along the front, with intervening spells of anticyclonic conditions. These are linked to the position and extent of the main air masses influencing the continental west margins in mid latitudes: the **Polar Continental**, **Polar Maritime**, **Arctic Maritime**, **Tropical Maritime** and **Tropical Continental** Air Masses. The interaction between these air masses, together with the associated upper jetstream and Rossby waves, influence the occurrence and development of depressions along the polar front.

Persistence of one of the continental air masses across these western margins can bring long spells of dry summer weather, but in winter 'anticyclonic gloom' conditions may occur. In contrast, the passage over the area of a deep, fast-moving depression can bring storm conditions with gale-force winds and heavy rainfall.

1.4 What are the causes of low pressure and high pressure hazards?

The role of jetstreams and Rossby waves in controlling the formation of weather systems

Jetstreams and **Rossby waves** control the formation of weather systems. Between the different atmospheric cells, at a height of about 5 miles, within the **tropopause**, are the jetstreams (Figure 13): the **polar jetstream** (40–60°N+S) and the **subtropical jetstream** (25–30°N+S). These jetstreams move air at high speed (up to 130mph) horizontally around the Earth and give rise to Rossby waves. The number of waves varies throughout the year but in summer there are usually between 4 and 6 waves. In winter, there are 3 waves.

In the mid latitudes of the northern hemisphere, Rossby waves are connected with the formation of **depressions** and **anticyclones**. As air travels west to east into a trough, it slows down and piles up and causes **convergence**. Convergence in the upper air causes a downflow to the ground, creating high-pressure systems at ground level. As air leaves the trough it speeds up and diverges ahead of the next trough. **Divergence** in the upper air causes low-pressure systems at ground level. At times the waves are few and shallow, giving a **high zonal index** and a succession of low-pressure systems. At other times the flow becomes more pronounced, giving a **low zonal index** and causing the formation of blocking, high-pressure systems.

Low-pressure system formation and associated hazards of storms, tropical cyclones and tornadoes

The definition of a climatic **hazard** is an 'extreme climatic/weather event causing harm and damage to people, property, infrastructure and land uses'. It includes not only the direct impacts of the climate/weather event itself but also the other (secondary) hazards 'triggered' by that event — e.g. landslides 'triggered' by torrential rain.

In the tropics, hazards associated with low-pressure systems are **tropical storms** and **cyclones** with torrential rain and high winds. These hazard conditions, usually created towards the end of the hot season (August–November in the Northern Hemisphere), are generated in exceptionally deep, fast-moving depressions over oceans off the east margins of continents in the tropics and subtropics. They trigger the secondary hazards of **flooding**, **storm surges** and **sea incursions**, **landslides**, **mudflows** and **windborne debris**.

In the temperate region, hazards associated with low-pressure systems include **severe storms**, **heavy rainfall** or **snowfall** and **gale-force winds**. These conditions are generated in exceptionally deep and fast-moving depressions, which are most likely to occur in autumn and spring along the polar front. They trigger the secondary hazards of flooding, sea incursions (especially where the deep depression coincides with a time of very high tides), landslides and windborne debris.

Tornadoes are small cells of very low atmospheric pressure formed where warm, damp air meets cool air from continental interiors. Tornadoes are associated with high wind speeds and cause structural damage to buildings along their narrow storm path.

High-pressure system formation and associated hazards of drought in tropical climates, or drought, frost and fog in temperate climates

In the **tropical** climates, the hazards associated with high-pressure systems are low rainfall, high evaporation rates and **drought**. These trigger the secondary hazards of **falling water tables**, **loss of vegetation**, **wildfires**, **soil erosion** and associated **desertification**. These hazards are associated with anticyclonic conditions, which are due to the continued persistence of the subtropical high pressure over continental areas. This limits the ITCZ zone to lower latitudes (nearer the Equator) than is normal for the time of the year. **Global warming** is a further contributory factor, exacerbated by people's misuse of their environment.

In **temperate** climates, the hazards associated with high-pressure systems include **drought** in summer and **frost** and **fog** in winter. They may trigger secondary hazards in summer: **falling water tables** and **loss of vegetation**; and in winter: **temperature inversion** with air pollution intensifying the fog conditions. These conditions are associated with a persistent stationary anticyclone, which in summer is usually associated with the extension into higher latitudes of the subtropical high pressure. In winter the conditions are usually associated with the extension of the continental high pressure towards the coastal margin of the landmasses.

1.5 What are the inter-relationships between human activity and climate?

The short-term and long-term effects of low-pressure climatic hazards on human activity

Low-pressure climatic hazards have both short-term and long-term effects on human activity. You should study these with reference to at least one specific low-pressure event in EITHER a tropical OR temperate climate (for an example, see Table 1).

Table 1 The impacts of Hurricane Katrina on the New Orleans area

Economic impacts	Social impacts	Environmental impacts
• Thirty offshore oil platforms damaged or destroyed and 9 refineries shut down. This reduced oil production by 25% for 6 months. • Forestry, port trade and grain handling severely affected. • Hundreds of thousands of residents left unemployed. Trickle-down effect with fewer taxes being paid. • Total economic impact in Louisiana and Mississippi estimated at over US$150 billion.	• Over a million people evacuated, displaced or made homeless. • Most major roads into and out of the city damaged as bridges collapsed. • Overhead power lines brought down by strong winds. Water and food supplies contaminated. • Worst-hit groups were those with no personal transport, less well off, non-white and vulnerable.	• Storm surge destroyed sections of the barrier islands and Gulf beaches. • 20% of wetland lost, affecting breeding of pelicans, turtles and fish. 16 National Wildlife Refuges damaged. • 5,300 km^2 of forest and woodland destroyed. • Flood waters containing sewage, heavy metals, pesticides and 24.6 million litres of oil pumped into Lake Pontchartrain.

The short-term and long-term effects of high-pressure climatic hazards on human activity

The hazards associated with high pressure have both short-term and long-term effects. You should study these with reference to at least one specific high-pressure event in EITHER a tropical OR a temperate climate. For example, you could study the effects of drought in Australia (for further information visit: **www.bom.gov.au/climate/drought/livedrought.shtml**).

The impacts of human activity on climate in both the short and long term

Here you have the opportunity to show the examiner your understanding of how human activity affects climate change (covered at AS, Unit G1).

1.6 What strategies are used to reduce the impact of climatic hazards?

Strategies to reduce the impact of hazards associated with low-pressure and high-pressure systems include **monitoring**, **prediction** and **warning** of future hazards, **immediate response** to lessen the impact once the hazard has occurred, and **long-term planning**.

Strategies to reduce the impact of low-pressure climatic hazards
In the case of Hurricane Katrina, which struck the Gulf coast of the USA in 2005, hurricane warnings were given and emergency services were in place but the strategies were not successful due to the failure of the levees to protect the city of New Orleans and due to the slow response of the US Federal Emergency Management Agency (FEMA).

Strategies to reduce the impact of high-pressure climatic hazards
In the case of drought in southeast Australia, sustainable solutions include water saving, water re-use and water treatment.

Strategies to reduce the impact of human activity on climate
Here you have the opportunity to show the examiner your understanding of strategies to reduce the impact of human activity on climate (covered at AS, Unit G1).

Theme 4 Development

1.1 What is development and what is the development gap?

Changing definitions of development
'Development' is very difficult to define, partly because the definition is dynamic. It is self-evident that different countries throughout the world are at different stages of development, but different groups of people mean different things by 'development'. A working definition is *'an increase in standards of living and quality of life for an increasing proportion of the population'*.

Earlier views on 'development' emphasised economic expansion and increased output, but the current view is much broader, involving social and cultural advancement as well as technological change and economic growth — and more recently, **sustainable development**.

Conventional development divides
Until the late 1980s countries were classified as belonging to either the 'First', 'Second' or 'Third' Worlds. The **First World** referred to countries that developed on the basis of capitalism, whereas the countries of the **Second World** developed on the basis of a command economy. The idea behind these terms was that each 'world' represented a path to development and **Third World** nations could choose between the first way (capitalism), a second way (communism), or invent a third way.

'**Developed**' is a term usually applied to a country or region that has a high standard of living and an advanced economy based on the effective utilisation of resources. Countries like this are often referred to as 'More Economically Developed Countries' (**MEDCs**), a term which acknowledges the fact that those countries that are not 'developed' in the sense of this definition may be developed in other non-economic ways, such as with regard to cultural, religious or social conditions. '**Developing**' is a term usually applied to a relatively poor country or region that has a low standard of living but which is beginning to achieve some economic and social development. In contrast to MEDCs, countries with low levels of economic development are referred to as 'Less Economically Developed Countries' (**LEDCs**).

The **Brandt Report**, published in 1980, highlighted the growing gap in social and economic development between the 'developed' countries of the world, **The North**, and the 'less-developed' countries, **The South**. The document was compiled by an independent group of statesmen headed by Willy Brandt, the former Chancellor of what was then West Germany. However, the North/South classification is simplistic and distorts a geographer's view of 'north' and 'south'.

The development gap
It is clear that there are groups of countries that share common characteristics. Some of these groupings are very polarised, leading to ideas of a **development gap**.

The development continuum
While it is true that massive contrasts occur, there are always countries that score at intermediate levels, leading some to conclude that there is a **development continuum** rather than a gap. The idea of a continuum tends to obscure the extent of extremes and many prefer to continue to use the term 'gap'.

1.2 How can development be measured and how useful are these measures?

Simple and composite indicators used to measure development
Many indicators may be used to quantify the level of 'development' of a country or region.
- **Simple** indicators measure only one aspect of development and include **GDP, GNP, GNI, adult literacy, life expectancy, daily calorie supply, infant mortality, cars per 1000 people, percentage of employment in agriculture, manufacturing and services** and even the **Big Mac index**.
- **Composite** indicators are more comprehensive as they measure more than one aspect of development. For example, the **Human Development Index (HDI)** gives a country a score from 0 to 1 using three variables based on adjusted income, education and life expectancy. On this basis Canada is the most developed nation, with an HDI of 0.971, and Niger is the least developed with an HDI of 0.340 (2009 statistics). The **Gender-Related Human Development Index (GRHDI)** is similar to the HDI, but adjusted for gender inequality.

Qualitative indicators
By **qualitative indicators** of development we mean aspects of development that may not be easily quantifiable. Qualitative indicators have been developed due to the recent emphasis on measuring development in terms of issues, such as **freedom**, **security** and **sustainability**, rather than by **statistics**. Qualitative indicators may be more problematic but reflect more accurately the ways in which development is now viewed.

The use of both **quantitative** and **qualitative** indicators is necessary in assessing the level of development of a given country and together they can yield important and unexpected insights into development that neither quantitative nor qualitative indicators could generate on their own.

Limitations of indicators

A problem with many indicators is that they can be misleading because most are averages and therefore do not show how far the benefits of development are shared within a population. Another problem with most measures of development is that they do not show the harmful side-effects that can occur. For instance, a rise in car ownership indicates a general rise in living standards economically, but some of the population will suffer from increased noise and air pollution and may take the view that their actual living standards have fallen. As with all statistics, development indicators are sometimes incomplete or inaccurate. Also, for some countries data are not reliable. Do collect examples of indicators that you can quote in the examination.

1.3 What factors have led to contemporary differences in development?

Physical, economic, social, political and cultural factors affecting the rate and nature of development

Frank's dependency theory suggests that developed countries control and exploit less-developed countries. This produces a relationship of dominance and dependency, possibly leading to poverty and underdevelopment in LEDCs. **Rostow** argues that all countries have the potential to break the cycle of poverty and develop through 5 stages of economic growth.

Within global regions, the level of development is often similar. Most European countries are well developed; most African countries are at low development levels and many parts of Asia are developing rapidly. But within regions of the world there are also variations dependent on resource endowment, government policy, and a wide range of other factors.

Malaysia has developed as a successful, second-generation, **newly industrialised country (NIC)** due to tax revenues from the **physical** resource of oil in the South China Sea, which provided initial funds for the country's development.

Political factors also contributed, including careful state planning with the government, led by Dr Mahatir Mohamad, managing the economy and controlling **foreign direct investment (FDI)**.

Economic factors were also significant, including not just the taxes from oil, but also low labour costs (11% of labour costs in the USA) and laws that favoured investors. Other contributory economic factors included the growth of export-led industries in economic priority zones (EPZs), free investment zones (FIZs), and free trade zones (FTZs).

Social factors, such as state investment in higher education, and **cultural** factors, including the 'Look East' policy and 2020 Vision (Malaysia hopes to achieve developed-world status by 2020) also help to explain the country's rapid rate of development.

The globalisation of economic activity and the rise of NICs/RICs and oil-rich countries

The opportunity to develop and the rate at which development has been taking place have been much influenced by the globalisation of the world economy. The biggest impact of this has been the **outsourcing** of manufacturing and services from developed countries to other parts of the world. This has, in turn, encouraged home-grown manufacturing in areas surrounding the focus of the outsourced activity. Tertiary activity has also moved out. Much of this has been low-level call-centre work, but higher-end activities, such as software design, have also moved.

Greater economic integration, such as that between Mexico and the USA, has stimulated development. Huge reserves of money generated by newly industrialised economies have made capital available to stimulate the establishment of new economic activities. The increased scale of the world economy has stimulated the extraction of raw materials and utilisation of energy sources, increasing their prices and injecting income into economies that had previously shown little sign of growth.

However, it would be wrong to class the entire developing world as exactly the same. Differences exist within the countries and between the countries.

Two significant groups of countries are the **oil-rich nations** and the **newly industrialised countries (NICs)**.

The 1970s saw huge increases in the price of oil as Arab countries withheld supply. This meant that any developing country with oil suddenly had a source of great wealth. Those without it faced a major disadvantage as they too had to pay the inflated price for fuel. The consequence for countries with oil was that they could now invest money in trying to alleviate poverty and promote education and health. Many also invested in industries that refined oil, adding great value to their export. Additional profits were invested in overseas ventures or loaned to other countries. As a consequence many of these countries, such as Saudi Arabia, amassed great wealth.

Newly industrialised countries are those that have recently seen substantial growth in their manufacturing output and, as a consequence, growth in their exports. They include South Korea, Singapore and Taiwan, as well as the territory of Hong Kong. Many have followed a similar system. Firstly, the country invests in industries that can produce goods that the country would normally import and supports these new industries by putting extra taxes on imported goods to make them uncompetitive. Then, when these industries are established, the countries look to replicate many of the products in the world export market. They concentrate on high-technology industries, first copying existing products and then improving them. The economy of such a country typically grows by 6–8% a year.

South Korea, for example, took advantage of its links with the USA. The USA, Japan and Europe provided the country with significant aid payments that it invested in iron, steel, shipbuilding, textiles and chemicals so it no longer had to import these. It also imported raw cotton and developed a textile industry. Once these were all established South Korea invested in industries that could provide export products such as computers, televisions and microwaves. Malaysia, India and China are examples of **recently industrialised countries (RICs)**.

1.4 How and why are development patterns changing?

Contemporary global development divisions

The World Bank classifies economies by their gross national income (GNI) per capita. There are 209 economies, classed as **low-income**, **lower-middle income**, **upper-middle income** or **high-income**. The classification is updated every year to keep pace with change. Of the 43 nations classified as low-income countries, 30 are located in sub-Saharan Africa.

Development issues in a changing world to include sustainable development and the status of women

The concept of **sustainable development** dates from the first Global Environmental Summit (1972), held in Stockholm. The concept was developed in the Bruntland Report (1987), where it was defined as *'development which meets the needs of the present without compromising the ability of future generations to meet their own needs'*. Although aspects of sustainable development can be measured, it is difficult to quantify sustainable development comprehensively, and also we do not know what the specific needs of future generations will be. However, for a country to be considered 'developed', it should be concerned about its environment and long-term sustainability. In this respect the USA cannot be considered developed in sustainable terms as it is responsible for the emission of 25% of the world's carbon.

Gender inequality is often a barrier to development. If only half the population can gain from development, any benefit is diluted across the whole of society, making the overall impact less. Countries with a low GRHDI tend to have their overall HDI depressed.

Economic change resulting in differentiation between different groups of countries

Economic change has affected world patterns of development, necessitating significant alterations to the 'three-worlds' model and Brandt's 'North/South divide' model. Some countries have declined or stagnated — such as Tajikistan in central Asia, since the breakup of the former Soviet Union, of which it was a member. Other countries have developed beyond all expectations, such as Malaysia. As a result the **Third World** or **South** has become substantially differentiated.

1.5 What hinders the closing of the development gap?

The burden of Third World debt
The greatest obstacle to development for an individual country is **indebtedness**. Countries that were at a low level of development in the past were loaned money through the World Bank and International Monetary Fund. Any money that the country generated had to be spent on paying interest on the loan before repaying the debt, and reinvestment in the economy was impossible. Such countries became caught in a poverty trap. They became what were defined as **'Heavily Indebted Poor Countries' (HIPCs)**. Special arrangements to relieve this debt have been developed by richer nations, such as the Multilateral Debt Relief Initiative (MDRI), but many believe this is still not enough to allow real development to take place.

Trade blocs
Countries have often come together to create trading blocs (e.g. the **European Union**), which greatly benefits each member of the bloc. However, countries outside the bloc may face quotas or tariffs that make it almost impossible to sell the commodities they have to offer. Within blocs, regulations often make it possible for producers, particularly of food crops, to generate huge surpluses. The surpluses are then sold at below cost price on world markets (dumping). Countries outside the blocs find that the commodities being 'dumped' are the same ones they themselves have to sell, but the dumped commodities are sold at below the price that could give the outsider countries any profit. This means the latter countries' only means to economic development is undermined.

Social constraints and cultural barriers
Gender inequality is a barrier to development. However, the **caste system** in India can also be viewed as a significant additional, cultural constraint. Hinduism is deeply rooted in India's culture, particularly through the caste system, which discriminates against the lowest caste, the **Untouchables** or *Dalits.* The Indian government has tried to reduce discrimination by ensuring that a percentage of public-sector jobs are reserved for certain subcastes of Dalits. Research in coastal villages in the Indian state of Andhra Pradesh showed that caste was a key factor influencing access to public services and facilities.

1.6 What types of strategies exist for reducing the development gap and how effective are these strategies?

Different types of aid: bilateral, multilateral and emergency aid
One way to try to overcome a low level of development is by accepting aid. However, not all aid is given with the intention of stimulating development. **Emergency aid** is directed towards providing relief after a disaster. One country may help another directly (**bilateral aid**) — often when a former colonial power assists a former colony after independence. More widely based assistance is usually termed **multilateral aid**. The aid may be monetary, but can be in the form of equipment,

education programmes and/or expertise. If such aid is specifically directed at promoting development it is often termed **structural aid**.

Many criticisms have been made of aid. Schumacher (1974) said aid was a process by which *'you collect money from poor people in rich countries and give it to rich people in poor countries'*. To be successful, aid must be effective, transparent and sustainable. Useful examples include education (particularly of girls), safe-water and sanitation projects, the provision of small loans (of less than US$200) to poor people, the eradication of diseases, and research to develop new crop varieties resistant to viruses and drought.

Free trade and fair trade
A point that has been noted is the way that economic blocs can hinder trade that could generate income and development. Removing barriers and promoting trade can allow development to take place. The **World Trade Organisation (WTO)** is committed to increasing free trade. However, the governments of countries with well developed economies are often reluctant to give up their position of advantage, and removing barriers is not as simple as it sounds. There are campaigns to promote **fair trade**, whereby producers are paid a reasonable price even when it is possible to drive the price down. Such trade encourages safe and non-exploitative working conditions, and avoids the use of child labour.

Foreign direct investment and the role of transnational corporations
Development has occurred extensively when manufacturing industry is introduced or opportunities to exploit valuable resources are taken. Governments of countries lacking development often work hard to encourage overseas-based companies to set up assembly or manufacturing plants or to mine resources. This may be arranged between governments but has most impact when a **transnational corporation (TNC)** is attracted and establishes a large plant or mine. Other companies that provide raw materials, services, transport and communications are also attracted and this often promotes rapid development.

The impact of TNC mining activity in Botswana has been very positive, with the country's per capita GNP increasing from US$800 in 1975 to US$16,500 in 2007. This has mostly been as a result of the mining of diamonds.

However, there can be a serious downside if a TNC moves on before a self-perpetuating multiplier has been established. Very often pollution occurs, and traditional attitudes and values come under great strain.

Debt rescheduling, debt abolition and debt-for-conservation swaps
As debt has been such a massive handicap to development, ways of eliminating debt have become important. **Rescheduling** can make repayment easier, but there is a strong lobby for **debt abolition**. The HIPC Initiative currently identifies 41 countries, most of them in sub-Saharan Africa, as potentially eligible to receive debt relief. One solution that also sets out to achieve a further objective is a **debt-for-conservation** arrangement. Here, a debt, or a portion of it, is written off and the money that would have been paid in interest on the debt is used to pay for conservation measures.

Such arrangements can be made by governments or by non-governmental organisations purchasing the debt to allow conservation. In Peru, ten of the most biologically diverse yet critically endangered rainforest areas escaped deforestation thanks to a landmark debt-for-conservation swap that the US Nature Conservancy helped facilitate in June 2002.

The UN has set targets for development through the eight 2015 **Millennium Development Goals**: *to eradicate extreme poverty and hunger, achieve universal primary education, promote gender equality, reduce child mortality, improve maternal health, combat HIV/AIDS, ensure environmental sustainability, and develop a global partnership for development.* The setting of these goals has had an excellent result in terms of highlighting issues, but success in achieving them has so far been limited.

One example of a narrowing of the development gap is that of Vietnam. Since the 1980s, FDI, improvements in trade — with membership of the Association of Southeast Asian Nations (ASEAN) bloc in 1995 and the WTO in 2006 — together with aid (e.g. £50 million a year from the UK Department for International Development) have operated to improve human development indicators and economic growth rates. Poverty fell from 58% in 1992 to 25% in 2005, life expectancy has improved, adult literacy is rising and primary-school enrolment rates are 97.5%. In the last 20 years, Vietnam reached an average economic growth rate of 7.5%. However, the country still faces challenges such as inequality, corruption, bureaucracy and environmental deterioration.

Theme 5 Globalisation

1.1 What is globalisation and global shift?

Concepts of cultural, economic, environmental and political globalisation

'Globalisation is the generic term for the process of integration in the realms of trade, economic relations and finance (it is broader, including social relations, knowledge culture and politics) and it is not new. It has been aided by the ICT revolution that has destroyed distance and indeed time. Brands are known the world over and are potentially destroying local diversity.' (Source: *Financial Times*)

The evolution of globalisation — stages in its development
The ultimate origins of globalisation could be dated back to the Roman Empire, but in the modern sense it evolved via colonialism and the growth of world trade and international financial systems. Its true modern origins lie in the 1975 **OPEC** oil price rises, when the new wealth of oil producers was invested in **MEDC** banks and loaned to developing countries and the emerging industrial economies that had cheaper labour costs.

Globalisation takes four forms:

1. **Economic** — the growth of TNCs at the expense of national governments.

2. **Environmental** — the creation of global problems that require global solutions.
3. **Cultural** — the increase in Western influence, especially American, over aspects such as music and the media.
4. **Political** — the increase in influence of Western democracies, the diffusion of state power to regional and international organisations such as the EU and UN, and an increase in the role of non-state actors, e.g. non-governmental organisations (NGOs) such as Save the Children.

The global shift as the movement of economic activities
Global shift involves the physical movement of economic activity from MEDCs, originally to **NICs** (newly industrialised countries), then to **RICs** (recently industrialised countries) and **LEDCs**. Initially the shift involved labour-intensive manufacturing, but increasingly it has involved all sorts of manufacturing and services, especially tourism.

1.2 What factors have led to current economic globalisation?
The factors behind the process of globalisation are outlined below.

Financial factors such as investment
Financial factors contribute, such as **foreign direct investment (FDI)**, where a company has at least a 10% interest in the investment in a receiving country. This investment has been made in order to lock into cheaper production costs (labour, raw materials), and cheaper operating and environmental costs. Another reason for investment overseas is that companies involved have sought to circumvent import restrictions such as **quotas** and **tariffs** on their goods. One reason why Nissan, a Japanese company, established a factory in Sunderland was to supply the European market with vehicles and thus avoid the payment of import duties into the EU. Several LEDCs have encouraged investment as a way of developing their economies.

Computer technologies
Computer technologies, such as broadband, the World-Wide Web, videoconferencing and email have speeded up the flow of information and communications. This has enabled business deals to be completed more efficiently and far more quickly.

Transport technologies
The reduction in the price and increase in the speed of transport technology have meant that goods and people can travel further, more cheaply and faster than at any time in history, and with ever-improving comfort and/or convenience. This has reduced the friction of distance and enables companies to locate more economically and take their product to the world market using extremely cheap and efficient transport modes. The tourism industry in particular has benefited from these factors.

The role of the World Trade Organisation
The **WTO** has been working towards promoting free trade between nations and reducing anti-competitive tariffs and quotas that restrict the integration and the flow of goods and services between countries.

Trade blocs

Trade blocs, e.g. the European Union, wield a lot of global power in trading matters. The very existence of trading blocs is a factor that is symptomatic of the process of globalisation.

1.3 How have companies globalised and shifted locations?

Global companies — TNCs/MNCs

The UN defines **TNCs** as corporations that *'possess and control means of production or services outside the country in which they were established'*. Their size is measured in terms of revenues, market capitalisation and, sometimes, employees. Most of the world's largest companies are American and include firms in both the manufacturing (e.g. General Electric) and service sectors (e.g. Wal-Mart stores). Significantly, the headquarters of TNCs are concentrated in Brandt's '**North**'.

The patterns of global manufacturing shift

In 1953, 95% of world manufacturing production was concentrated in the industrialised countries. Although manufacturing remains concentrated in North America, western Europe and Japan, decentralisation has occurred as a result of investment by TNCs in the three generations of NICs in the developing world. NICs have created large companies of their own that are now locating factories in developed countries like the United Kingdom.

Location factors for the global shift

Location factors that influence the global shift include the availability of a large, disciplined and skilled workforce, suitable infrastructure, political stability, government incentives and a large domestic market.

Service-sector shifts

Outsourcing or **offshoring** is the global shift of services from MEDCs to NICs, RICs and LEDCs. India and countries of the Indian Ocean rim were forecast to receive 2 million jobs from MEDCs by 2008. India has become a common destination for outsourcing due to: Commonwealth links with the UK, high levels of English-language skills, strength of IT education, lower communication costs, and lower wage and capital costs.

The impact of outsourcing and offshoring

Outsourcing and offshoring can bring considerable benefits for countries like India in terms of job creation, higher salaries, greater disposable incomes and a reduction in gender apartheid. However, there are also disadvantages including Westernisation and consequent loss of cultural identity, unsocial hours, and increasing social divisions.

The impact of outsourcing and offshoring for MEDCs is simple: more profitable returns for the participating companies. This enables them to maintain employment in the **quaternary** jobs in their home country and in the manufacturing/service jobs in the production countries. These advantages must be set against significant job losses in the service sector in MEDCs, particularly jobs typically held by women in vulnerable **deindustrialised** areas.

1.4 Who wins from the global shift and globalisation?

Global development indicators that identify NICs and RICs

NICs are countries where industrial production has grown sufficiently so that it becomes a major source of national income. Among the first-generation NICs were Hong Kong, Taiwan, South Korea and Singapore. The list of countries that can be called NICs has grown over the past 40 years.

'Recently industrialised country', or RIC, is a term used to cover those countries that have tried to emulate the first-generation NICs. Such countries include Vietnam, Indonesia, Chile, China and India.

GNP per capita in US dollars is the sum of all goods and services produced in a country plus taxes and income from abroad, divided by the population. NICs such as Singapore and Hong Kong have GNP figures that approach those of MEDCs.

The rise of the NICs/Asian Tigers

Malaysia has developed as a successful second-generation NIC due to:
- Careful state planning, with the government managing the economy and controlling FDI.
- A 'visionary' leader, Dr Mahatir, who led a single-party government for two decades; it had strong controls on the media.
- Taxes from oil in the South China Sea, which provided initial funds for development.
- Low labour costs and laws that favoured investors.
- The growth of export-led industries in economic priority zones (EPZs), free investment zones (FIZs) and free trade zones (FTZs).

Benefits of being an NIC, and benefits to investing countries

Economic benefits for NICs include the expansion of industries and services, increased international trade, rising incomes and infrastructure improvements. Social benefits include an expansion in employment opportunities. Environmental benefits may include the development of ecotourism, aided habitat preservation, and national park development. The benefits to investing countries include efficiency gains and improved profits with lower costs.

The rising superpowers: India and China

Brazil, Russia, India and China together form a group known as 'the **BRICs**' — the fast-growing developing economies. Of these, India and China are seen as emerging superpowers. India's rapid economic growth has been due more to expansion in the service sector than in manufacturing. China's accelerating growth is due mainly to the expansion of manufacturing fuelled by foreign direct investment from Japan, the USA and Europe.

1.5 Who loses from the global shift and globalisation?

The negative effects of being an NIC socially and environmentally

- Environmental degradation results from the exploitation of primary resources, e.g. the Carajás iron ore project in Brazil.

- Deforestation has occurred on a large scale. A particular problem is 'the haze', a smog caused by burning forest as it is cleared, e.g. Indonesia.
- Pollution occurs as controls are less stringent: the lack of adequate controls led to the 1984 Bhopal gas-leak disaster in India that caused 22,000 deaths.
- There is a greater divergence in earnings.
- The working conditions of the labour force are often reported to be unhygienic, with very long hours, lack of union representation, no sick pay and none of the social benefits enjoyed in MEDCs.
- More women are employed, which can reduce the birth rate to below replacement level.
- Use of expatriate skilled workers, especially in the finance sector (e.g. in Singapore), and in unskilled sectors such as construction can mean that distinct, 'ghettoised' expatriate and immigrant areas develop.
- The population becomes more Westernised, contributing to a lack of cultural identity.

Factors leading to deindustrialisation

Deindustrialisation is the decline in manufacturing that has been experienced in the regions of North America and Europe that industrialised in the nineteenth and early twentieth centuries. This has come about for a number of reasons:
- Many products are at the end of their life cycle — newer products have replaced the old.
- Outmoded production methods have been replaced by newer technologies requiring less labour.
- Labour has clung to old practices and there has been poor labour management.
- There is competition from cheaper locations with low labour costs.
- There was a recession in the 1980s.
- Government support for industries such as coal and steel has been removed.
- Firms have needed to rationalise production.

Changing employment in MEDCs

Responses aimed at reviving regional economies suffering from deindustrialisation may include promoting location, developing leading industries, creating research and development (R&D), providing government assistance at the local, regional and national level, and encouraging tourism. These measures have resulted in the growing **tertiarisation** of economic activity. Deindustrialised areas of the UK include south Wales and north-east England.

The environmental effects of globalisation

Globalisation has negative effects at global level, such as global warming, and these require global solutions (covered at AS, Unit G1).

Globalisation can also have negative effects at regional level — for example, the exploitative nature of TNCs can lead to problems, as with the Bhopal disaster involving the US-based chemical firm Union Carbide, and the deforestation of vast areas for the purposes of oil-palm cultivation that has created 'the haze' in Indonesia.

1.6 What are the causes and effects of political and cultural globalisation?

Empires and superpower status
Globalisation has led to a situation where most countries are interlinked in various ways — politically, culturally and economically. Causes of political globalisation include the influence of the **superpowers**, particularly the USA. The superpowers are often criticised for exploiting the situation of lesser-developed countries, as in the case of US attempts to exert pressure on the Iranian government to stop developing nuclear technology, which may be seen as infringing Iran's sovereignty.

Westernisation and cultural integration
Cultural integration refers to the increased exchanges of cultural practices between nations. New technologies, such as commercial air travel, satellite television, mass telecommunications and the internet, have created a world where billions now consume identical cultural products, e.g. pop music. People also follow similar cultural practices, such as eating the same foreign food and adopting the same foreign words. The operations of TNCs have resulted in a product and lifestyle monoculture. Effects of cultural globalisation include reduction in cultural diversity, the loss of cultural identity and the development of a **Westernised** consumer culture.

The rise and re-emergence of other cultures
A rise in **nationalism** and **fundamentalism** has occurred in response to the process of globalisation, with many countries and regions making an attempt to retain their own cultural identities.

Globalisation and the development gap
One of the key negative effects of globalisation has been a growing **development gap**, with increasing disparity in levels of development between countries. This has been fuelled by the uneven pace of development around the world. For example, trade patterns and patterns of foreign direct investment reveal the special economic problems of sub-Saharan African countries.

Theme 6(a) Emerging Asia: China

1.1 What are the main physical and demographic characteristics of the country of China?
The People's Republic of China is located in east Asia, on the west side of the Pacific Ocean. It is the third-largest country in the world after Russia and Canada.

A brief overview at the national scale of patterns
(i) Climate
Although most of China lies in the temperate belt, its climatic patterns are complex, ranging from **subtropical** in the south to **sub-arctic** in the north. **Monsoon** winds

dominate the climate and have a major influence on the timing of the rainy season and the amount of rainfall. Alternating seasonal air-mass movements and accompanying winds produce moist summers and dry winters.

(ii) Relief, drainage and water availability
China's relief is both complex and variable, ranging from mostly mountains, high plateaux and deserts in the west, to plains, deltas and hills in the east. The Qinling Mountains provide a natural boundary between north and south China.

The Tibetan Plateau in the west is the source of almost 50% of the major river systems in China, including the three longest rivers: the Yangtze, Huang He (Yellow) and Pearl rivers. These flow west to east, into the Pacific Ocean. About 10% of Chinese river systems drain into the Indian or Arctic oceans. The remaining 40% have no outlet to the sea; they drain through the dry western and northern areas of China, forming deep underground water reserves.

(iii) Natural resources
China has a range of natural mineral resources including coal, iron ore, petroleum, natural gas, mercury, tin, tungsten, antimony, manganese, molybdenum, vanadium, magnetite, aluminium, lead, zinc and uranium. China has the world's largest hydro-power potential, with reserves of 680 million kw.

(iv) Population distribution
China's population of 1.3 billion is the largest in the world. It is concentrated in the eastern and coastal part of the country and along major rivers such as the Yangtze, Huang He and Huai. Much of the country, including the steep Himalayas, the dry grasslands in the north, the central region and the Gobi Desert in the north, is almost uninhabited. Nearly 60% of Chinese live in rural areas.

(v) Regional differences in levels of development
Economic growth has taken precedence over equality as the main development goal since the economic reforms of the late 1970s. In the 1980s, under the slogan of 'letting some people and some regions get rich first', China implemented a coastal development strategy. Now, the prosperous urban coastal zone is in sharp contrast to the poor, rural interior.

1.2 Why and how is the economy changing?

Changes in economic policies
After the death of Mao Tse-Tung (Mao Zedong) in 1976, China's economy took a major change in direction. In 1978, Deng Xiaoping, the new leader of the Chinese Communist Party, introduced the 'Open Door' policy, which was designed to overcome China's isolation from the world's economies. The country had become increasingly aware that the world, and south-east Asia in particular, was developing and leaving China behind. China moved towards a **socialist market economy**. Today, China's twenty-first century leaders are able, educated, and have less ideological baggage than those of the 1980s. They are focused and determined on economic growth at all costs — but on China's terms.

New industries in the changing economy
Between 1949 and the late 1970s manufacturing in China was undertaken almost entirely by **state-owned enterprises (SOEs)**. These were mainly heavy industries such as oil, chemicals, power, iron and steel. The 1980s focus on increased productivity forced SOEs towards reform. Large SOEs have improved their management and smaller SOEs eventually privatised. Chinese firms have gradually become more like Western companies.

There has been some success for large SOEs. Industrial output has grown by up to 13% per year and there have been major improvements in terms of technology, managerial skills and efficiency. SOEs have attracted **TNCs** as partners and **FDI** (foreign direct investment) has been highly significant. FDI in China increased from US$3.5 billion in 1990 to US$93.7 billion in 2007.

Factors affecting the growth of new industries and the contrast between coastal areas and the interior
Since 1979, five **special economic zones (SEZs)** and 14 **open cities** have been proclaimed. These offer reduced restrictions on land, labour, wages, taxes and planning regulations to overseas firms, especially those involved in high-technology industries. The result has been the emergence and dominance of economic activity in coastal areas — which have received most of the internal investment as well as having imported capital, technology and entrepreneurial skills, at the expense of the interior.

Impact of the changing age structure on the economy
A **positive demographic dividend**, where there are more people of productive age with a low dependency ratio, has coincided with the economic boom. Output per capita will rise by 10% between 1982 and 2050, but although China's economy is growing it is not becoming more efficient. The **negative demographic dividend**, associated with the rapid process of ageing, will soon affect China. As people live longer they have to either accumulate wealth or face a reduction in their standard of living in their old age.

1.3 What are the economic and social challenges facing rural communities?

Changes in the organisation of agriculture and rural economic activities
The rural economy was the first sector to be reformed because low agricultural productivity posed problems of food security for China. In 1949, all land was brought into communal ownership under the guidance of the state. There was initially an increase in food production as a result of **Green Revolution** farming methods, but this was not sustained. In 1981 farmland was divided up between households, with every household given a 15-year contract to farm the land. This security gave farmers confidence to invest and to manage land more effectively. The reforms revitalised the rural economy. Since that time, agricultural production has increased very slowly and China is beginning to experience food insecurity again.

During Mao's era, rural industries called **town and village enterprises (TVEs)** produced heavy goods such as iron, steel, cement, chemical fertiliser and farm tools, as well as hydroelectric power. After 1978 these enterprises expanded so as to develop a wider range of businesses. Many Chinese farmers preferred to invest their resources in rural industry rather than agriculture. This encouraged the growth of small businesses run by the most successful peasants. Thus a new entrepreneurial class began to emerge and TVEs have become the backbone of development in rural areas.

The effect of population policies in rural areas
The Chinese government's **one-child policy** has been resisted in rural areas, due to the need for labour. The 1980 rural reforms encouraged rural families to demand more family labour and some provinces have conditions for exemption from the one-child policy.

The impacts on and challenges of migration for rural areas
Before the 1980s internal migration was tightly controlled in a bid to avoid unmanageable growth in cities. A household registration system called *hukou* controlled access to basic requirements such as housing, welfare and employment, and was not transferable between districts. Once one was registered with a rural hukou the registration would be permanent. However, since the 1990s demand in cities for unskilled and semi-skilled labour has led to rural migrants being given a temporary urban hukou, which enables them to have access to some housing and basic welfare in cities. The registration requires annual renewal, so that the authorities can control the number of migrants. Temporary hukou sets the migrant population apart and has increased inequality of access to services in cities.

Social welfare services such as health and education
There is a significant divide between rural and urban populations, reinforced during the Mao years by the hukou system. Progress and development in urban areas was not matched in rural areas. Consequently, many rural areas are extremely backward, traditional, have very poor services and amenities, and essentially are a world apart from the modern regional and provincial cities. Rural education and health facilities are poor, particularly for an aspiring superpower like China. Villagers often lack any form of social safety-net such as pensions or health insurance.

Sustainable development
In many rural communities the focus on economic growth and new industries is putting pressure on the environment. Deforestation, air and water pollution and the conversion of land from agricultural to industrial use are gradually putting food production in jeopardy. Increased pressure on the remaining farmland increases the risk of soil degradation. Villages and small towns have to increase their own incomes, mainly through small industries, if they are to contribute to health and education services. If these services decline, outmigration will increase and communities will have even greater difficulty in developing businesses and maintaining basic services.

1.4 What are the economic and social challenges facing urban communities?

Changes in the organisation of economic activities in urban areas

Each urban area competes for a share of economic growth and foreign investment. High-technology industrial parks, tariff-free districts and **SEZs** aim to attract new businesses, usually located away from existing, crowded and fully developed urban centres. In some cases, provincial and local governments speculate on development land. The result is thousands of hectares of empty plots of land surrounding one or two high-rise office buildings on the city edges. There is often a lack of coordination between central and local authorities. Some communities are keen to sell their land to developers so that they can change their hukou designation.

Migration to urban areas and increasing social inequality

Millions of people have migrated from rural to urban areas to fill the jobs generated by the economic explosion. Most migrants head for the eastern-seaboard cities of Beijing, Tianjin, Tangshan, Shanghai, Changjiang and the Zhujiang delta. There is also growth in provincial capitals. However, anti-poverty campaigners argue that many workers receive low wages and live in poor conditions. Every year an estimated 200,000 people move to slums on the southern outskirts of the capital, Beijing. Social inequality is increasing between the growing middle class in urban areas and the poor and migrant population. Access to welfare, health and education provision is much better for those residents with permanent urban residential hukou status.

Social welfare services such as health, education and housing

Social challenges associated with China's rapid urbanisation include deprivation and poverty, segregation, and problems associated with health and crime. Housing inequality reflects many of the welfare problems facing urban China.

Increasing rural–urban inequalities

Capital-intensive urban development has created a large productivity gap between the agricultural sector and other sectors of the economy. This has led to a growing gap between rural and urban income per capita. Migration from rural to urban areas, to fill the jobs generated by the economic explosion, has further increased these inequalities. In 2000, the per capita GDP of Shanghai was nearly 10 times that of Guizhou province in south-central China.

Sustainable development in towns and cities

As towns and cities grow sustainable development becomes more challenging. Demand for energy is rising. People and industries demand more water, so supplies in lakes and groundwater reservoirs fall. As land at the edge of cities is developed for new factories there is less available for farming. City sprawl separates homes from industries, increases the amount of commuting and creates traffic congestion. Domestic and industrial waste disposal facilities come under pressure, particularly from the emerging middle classes. National, provincial and city authorities are not always willing to pay for sustainable projects and services.

1.5 What are the effects of globalisation on China?
The role of foreign firms in changing and developing the economy
The liberalisation of trade since the 1980s has led global TNCs to expand aggressively in search of new emerging markets. By encouraging foreign firms into China, competition has raised levels of efficiency and forced large SOEs either to modernise or to close down. **Joint ventures (JVs)** have been vitally important for China, with firms such as Procter & Gamble, Caterpillar and United Technologies being particularly successful. Key features of JVs have been the requirement for technology transfer and an insistence that subcontracted work is given to selected domestic firms. This ensures that China acquires 'know-how' which can then be transferred to domestic firms.

The importance of exports and the role of the WTO
China's export 'basket' consists of labour-intensive export products such as toys, clothes and assembled electronics, as well as more sophisticated products that are more typical of a country with a much higher GDP per capita. The JVs, located in clusters in SEZs, provide a critical source of technology and technology transfer and they dominate exports. China's membership of the WTO since 2001 continues to be a driving force in the opening up of China to both imports and exports. This will have wide-ranging impacts on economic and political systems in China, particularly on the ways in which business is conducted.

The economic and political impacts of China's trade with the rest of the world
China needs resources for its continued economic growth and has been determined to establish trading relationships with countries that can supply raw materials. The increase in food and mineral imports into China has had the effect of driving up many world commodity prices, such as those for iron and other ores. China has increased trade and foreign direct investment on all continents, including into Europe and the USA, and there have been periodic rows with China's trading partners concerned about 'dumping' of exports.

China has always traded with Africa, but recently international attention is being focused on that trade and the political influence that accompanies it.

Within ASEAN (the Association of Southeast Asian Nations) there is some fear about the influx of Chinese goods into local markets. Countries cannot compete with China and therefore must find alternative strategies in order to sustain what economic development they have already achieved.

1.6 What are the environmental challenges and solutions facing China?
The causes and consequences of soil erosion, industrial pollution, sustainable use of water resources, and the need for energy supplies
It is estimated that since 1949 China has lost one-fifth of its agricultural land to soil erosion and economic development. Air pollution is a major issue in cities such as Beijing and Shanghai due to their heavy reliance on coal. Water shortages are also a

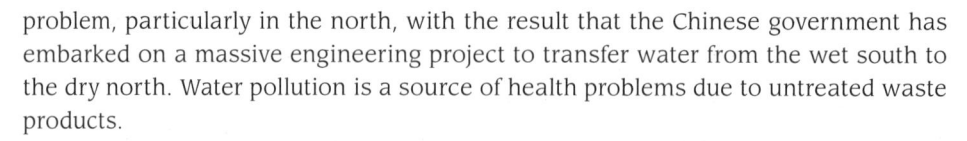

problem, particularly in the north, with the result that the Chinese government has embarked on a massive engineering project to transfer water from the wet south to the dry north. Water pollution is a source of health problems due to untreated waste products.

The balance between economic growth and sustainable development
There is growing environmental awareness among grassroots organisations and communities in China, but serious concern for environmental sustainability within the Politburo (the Communist Party's governing body) is still overridden by the desire for economic growth. Despite that, the government response to the Rio and Kyoto agreements on the environment suggested some recognition of the need for sustainability. China signed the Kyoto Protocol in 1998, less than a year after it was set up. This was also intended to establish China as a leader among the developing nations. Environmental concerns are being taken seriously, but bureaucratic problems and some corruption inhibit the implementation of national policies at local level.

Theme 6(b) Emerging Asia: India

1.1 What are the main physical and demographic characteristics of the country of India?

A brief overview at the national scale of patterns

(i) Climate
It is not easy to generalise about India's climate as the country covers such a large area — India makes up the majority of the Indian subcontinent — and its climate is strongly influenced by both the Himalayas and the Thar Desert. India has six climatic subtypes ranging from desert to alpine tundra. In general, temperatures tend to be cooler in the north, especially between September and March. India has four seasons: winter (January and February), summer (March to May), the wet **monsoon** season (June to September) and the dry monsoon season (October to December).

The wet monsoon season occurs with the movement of the **ITCZ** into the region bringing an area of low pressure and drawing in hot, moist winds from the ocean. Rainfall is increased by **orographic uplift** where these moist winds are drawn over uplands such as the Western Ghats. Temperatures average 30°C and humidity is also very high, with average rainfall around 2000 mm, decreasing with distance inland. Cyclones and hurricanes are frequent towards the end of the rainy season (Figure 15, p. 34).

The cooler dry season coincides with the extension of continental high pressure as the ITCZ moves back towards the Equator and across into the tropics beyond. With high pressure dominating, there is air subsidence and outblowing winds are dry.

Temperatures remain relatively high at 25°C in lowland areas and evaporation rates are also high. The weather is much more severe in mountain areas (Figure 16, p. 35).

(ii) Relief, drainage and water availability

The major rivers of India originate in one of three main watersheds: the Himalaya and Karakoram ranges in the north; the Vindhya and Satpura ranges in the centre; and the Western Ghats in the west. The Himalayan river networks are snow-fed and have flow continuously, throughout the year. The other two networks are dependent on the monsoons and have significantly lower discharges during the dry season.

(iii) Natural resources

India's major mineral resources include coal (India has the third-largest reserves in the world), iron ore, manganese, mica, bauxite, titanium ore, natural gas, diamonds, petroleum, limestone and thorium (the world's largest deposits are in Kerala, in the south). Oilfields off Mumbai and onshore in Assam meet 30% of the country's demand; however, India is still heavily dependent on imports of both coal and oil for the rest of its energy needs.

(iv) Population distribution

India's population — 1.1 billion in 2008 — is the fastest-growing in the world and by 2030 India's population will have permanently overtaken China's as the world's largest. The population is concentrated in the fertile northern floodplains. Four states — Bihar, Madhya Pradesh, Rajasthan and Uttar Pradesh — account for 40% of the population and 47% of population growth. The population of Uttar Pradesh, 166 million in 2001, is roughly the same size as that of Pakistan and of Bangladesh. The southern states have lower fertility rates than the north, a contrast which will become more and more marked.

(v) Differences in development between states

India is organised as a federation of states and union territories, each of which has considerable political independence. Since 1991, economic growth in India has been characterised by increased inequality between states — average income in Punjab, Gujarat, and Maharashtra is four times that of Bihar.

1.2 Why and how is the economy changing?

India has undergone major changes in economic policy. After independence and Partition (whereby British India separated into India and Pakistan) in 1947, India's aim was to develop economically without the participation or influence of foreign capital. Economic policies had a strong anti-export bias. Socialist governments ensured a high level of state control over key industries, which in turn led to excessive bureaucracy and very slow economic growth.

A major economic crisis in 1991 forced the governing Congress Party to borrow money from the **International Monetary Fund (IMF)**. This opened up the economy to economic globalisation. India is now among the ten fastest-growing economies in the world.

Changes in traditional agriculture

Agriculture in India is characterised by unequal productivity across the country. The highest yields are found in Punjab and Haryana and the lowest in the northeastern states of Bihar and Orissa. In the 1960s India was highly dependent on imported food. However, the **Green Revolution** had a major impact on Indian food production. The area under HYV (high-yielding variety) wheat crops increased from 4 hectares in 1963 to 4 million hectares in 1971. The wheat revolution was followed by similar changes involving rice, sugar, millet and oilseed crops, as well as cotton. As a result, food production exceeded population growth. The Green Revolution was successful in raising incomes for farmers on naturally fertile soil, but increased inequalities between wealthy farmers on productive land and poorer farmers on marginal land.

The role of agribusiness

Agribusinesses play an increasingly significant role in agricultural exports and in food security. They control much of the chain, from seeds and fertilisers to finance, distribution and marketing. The increased output from new, large farms in India helps to maintain food security.

The growth of service and financial industries

An extensive financial and banking sector supports the rapidly expanding Indian economy. India has a wide and sophisticated banking network. The sector includes a number of national and state-level financial institutions and a well established stock market. The Indian capital markets are rapidly moving towards a modern market including derivative and internet-based trading. The services sector, including financial services, software services, accounting services and entertainment industries like Bollywood, contributes 41% to GDP.

Factors affecting the growth of manufacturing industries

In addition to political changes, factors responsible for the rapid growth of manufacturing industries include economic change, such as the emergence and investment policies of **TNCs**. Other contributing factors are the growth in Indian firms and of an urban, educated, middle-class population whose members have become consumers themselves and who provide a large market for new consumer goods. Technological factors have also played a significant role, particularly the speed and distance over which communications and movement can now take place due to changes in computer, transport and communication technologies.

The need for major developments in infrastructure throughout India

India has serious transport issues. The road transport sector has been declared a priority and will have access to loans at favourable conditions. The country needs US$200 billion worth of new ports, roads and other infrastructure and another US$50 billion to modernise 40,000 km of roads. Currently there are significant delays in distribution and severe bottlenecks, with state and federal governments often in opposition. The National Highways Act has been modified to help reduce tolls on national motorways, bridges and tunnels. The government is also implementing a new policy that aims to improve India's telecommunication systems.

1.3 What are the economic and social challenges facing rural communities?

The traditional socio-economic characteristics of rural India

'*India lives in its villages*' is a quote from the Registrar-General (2005). Although our impression of India is of overcrowded cities, in 2001 the average Indian lived in a settlement of 4,200 people. Some 72% of the population is classed as rural, with 58% being farmers. While in some ways India is rapidly becoming a middle-class country with Western lifestyles, in other ways it remains a rural country where social and religious traditions are embedded.

Most rural Indians have lower educational levels, higher mortality and fertility, greater poverty, and access to fewer services and amenities than urban dwellers. Most Indians live their whole lives in a relatively limited geographical area. Some rural areas in the states of Bihar, Jharkhand, Uttar Pradesh and Orissa are officially destitute.

The impacts of migration

Migration to cities has occurred due to a lack of opportunities in rural areas. This push factor affects the sustainability of urban growth, which has become a concern for state and national governments. Development of rural areas could do much to stem internal migration and take pressure from urban centres.

Social welfare services in rural areas

There are many welfare concerns, such as the need to provide minimal social and income security for agricultural workers. Education is a challenge in many rural areas, particularly the education of girls, and dropout rates are high and attendance is poor. Poorer agricultural households show the worst attendance levels, especially in migration and harvest seasons.

Food production and combating hunger after the Green Revolution

Economic challenges associated with India's rural communities include the challenges of food production and land reform and problems with infrastructure and service provision. **Social challenges** include population growth, hunger, malnutrition and migration.

The relationship between population growth, hunger and sustainability in rural India

India's growing population increases the challenge of food security. There is growing demand for fertile farmland to be used by multinational companies to grow industrial and food crops for export. New industries also demand land in order to expand. Both of these developments also use up water resources and increase pollution of soil and water.

Poor farmers are often forced onto more marginal land that, without expensive fertilisers and pesticides, produces lower yields. Those farmers who do try new technologies are at risk of debt if crops should fail. With more people to feed and less quality land available, the poor face an increased risk of hunger.

A key question for India is how to provide sufficient food for an expanding population while at the same time encouraging agricultural and industrial development. Individual states have very different attitudes to this problem.

1.4 What are the economic and social challenges facing urban communities?

The rapidly growing and economically powerful urban areas contrast with traditional and often remote villages. Only 11% of India's population live in cities of over 1 million people. The number of city dwellers in India increased by 60 million between 1991 and 2001, but the number of rural dwellers increased by 113 million. There is a crisis in urban infrastructure due to that extra 60 million, but also due to the aspirations on the part of the middle classes for private cars and a higher-consumption lifestyle.

Changes in type of economic activity
Indian cities with over 1 million people are integrated into the global economy but smaller cities tend to look towards the local economy.

Migration to urban areas and the interdependence of rural and urban populations
In India, rural–urban links are strong. Journeys of 20–30 km into Mumbai, Kolkata and Chennai are not uncommon. Family members return to rural homes to help with the harvest, and money sent by urban dwellers to rural areas makes a significant contribution to poverty reduction despite increasing urban poverty. The reverse also happens, with rural families sending money to urban relatives in order to support students or men seeking work.

Delivering modern infrastructure and social welfare services
Social challenges include deprivation and poverty, segregation, and problems associated with housing, health and crime. As a result, most urban settlements are characterised by shortfalls in housing, inadequate sewage, poverty, and social unrest. This makes urban governance a difficult task.

Increasing inequalities within urban areas; the informal sector and urban poverty
The poor compete with middle-class people for land, which may be wanted for building homes, transport installations and retail malls. Economic challenges associated with India's urbanisation include the growth of the informal sector, problems of service provision, and exploitation of the labour force.

Sustainable development in cities
Five Indian cities are included in the global Sustainable Cities Programme: Chennai, Bangalore, Hyderabad, Delhi and Kolkata. The key development issues for the

sustainable development of cities are to secure housing rights, to provide access to civic amenities, public health, and education, to provide safe and secure drinking water, and to improve food security. In addition, there needs to be freedom from violence and intimidation and the provision of adequate social-security programmes.

1.5 What are the effects of globalisation on India?

The impacts of global trade on the national economy

Since 1991, the economy has undergone a major transformation. The high levels of protectionism have been replaced by growth in exports. India is building economic and political ties around the world. It has considerable influence over world trade as a founder signatory of the General Agreement on Tariffs and Trade (GATT), the forerunner of the WTO. India leads the developing nations in global trade negotiations and is trying to encourage a more liberal global trade regime, especially in terms of services. India is one of the top ten exporters of services in the world and is famed for its specialist trade in IT services. The direction of trade is also changing, away from Russia and eastern Europe towards the USA, EU and east Asia. India's major trading partners are the USA and China, but it is also developing trade links with African countries.

The effects of the 1991 debt crisis and structural adjustment

As noted, a major economic crisis in 1991 forced the governing Congress Party to borrow money from the International Monetary Fund. This triggered a major change in the economy, allowing direct foreign investments into the country, which opened India up to economic globalisation.

The growth of Indian TNCs

Some industries, such as defence and aerospace, remain under state control, but many manufacturing sectors, including vehicle, consumer electronics and white-goods manufacturing, are now open to foreign direct investment. Indian companies may set up joint ventures or become wholly-owned subsidiaries of foreign firms.

Conflicting views of the benefits of globalisation for India

Beneficial effects of globalisation on India include foreign investments into pharmaceutical, petroleum and manufacturing industries, which have provided a significant boost to the Indian economy along with new employment opportunities, and have contributed to the reduction in levels of unemployment and poverty. Foreign companies bring advanced technology, helping to make Indian industry more technologically advanced. As a result India has seen an increase in international trade with a growth in exports, rising incomes, and infrastructure improvements.

However, globalisation has also had some negative effects, such as increased competition in the Indian market between foreign and domestic companies. Other negative effects are growing consumer preferences for buying foreign goods, and a reduction in labour requirements due to the introduction of advanced technologies that require less labour — especially in heavy manufacturing.

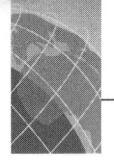

The impact of globalisation on India's poor

Globalisation has led to widening social and regional disparities. India's economic development, particularly since 1991, has increased inequality between castes and between states. Although India has developed in terms of rapid industrialisation, consumerism and materialism, in many cases this 'progress' has not affected everyone. In fact, many people are worse off than in 1991 — especially the *Adivasi* (the scheduled tribes and castes and indigenous communities), the landless peasants, and marginalised farmers. The real challenge for India is how to enable the benefits of economic growth to trickle down to the very poor. If India continues to grow at 8% per year, average rates of poverty will fall to single figures within 20 years. This would be a significant achievement but would still mask many underlying economic inequalities across the nation.

1.6 What are the environmental challenges and solutions facing India?

The causes and consequences of deforestation, soil erosion, industrial pollution in major cities, sustainable use of water resources, and the need for energy supplies

After independence India's forests were exploited commercially for pulp and paper, and they continue to be overexploited. Huge areas have been deforested, accompanied by serious soil erosion. There has been significant replanting by state forestry departments, but this has been primarily of commercial timber — eucalyptus, teak and pine. Although the forested area is increasing, the loss of biodiversity is not sustainable.

The Ganga (Ganges) River, which has great spiritual and emotional significance for Indians, is seriously polluted. In 1984, the Department for the Environment set up an action plan to reduce pollution. Initiated in 1986, the plan aimed to intercept, divert and treat 882 million litres of waste water per day.

India has both too much and too little water. The areas that flood in the monsoons, and the number of people that are affected, increase every year. Increased demand in rural areas results from the demands of agriculture, rising standards of living and rural industrialisation. The unpredictability of rainfall in many semi-arid areas is increasing because of failed monsoons and climate change (covered at AS, Unit G1).

India faces a critical challenge to meet a rapidly increasing demand for energy. India is sixth in the world in terms of energy demand. There are significant reserves of coal, but limited oil and gas resources. However, India has considerable potential for exploitation of renewable energy resources such as solar and wind power, and biofuels (from sugarcane).

The balance between economic growth and sustainable development

The challenges of addressing poverty as well as managing the environment sustainably remain significant. After the UN Conference on Human Environment in 1972, India created a National Committee on Environmental Planning and Coordination

(NCEPC). Environmental issues were then included in the national 5-year plans. In the 1980s the Ministry of Environment and Forests was created. Now, numerous autonomous agencies, offices and institutions exist, set up by the government at national and state level.

There is a will to improve the environment, but this often conflicts with other demands and, as in most other countries, sets government departments at odds with each other.

India faces many of the same environmental issues as developing countries. It is challenged by the need to meet the demands of industrialisation for development while understanding the necessity for environmental sustainability. The rates of urban and rural change make it hard to ensure that the best environmental decisions are taken. India's democracy can hinder progress. The sheer scale of the environmental challenges is daunting, yet progress is being made at national and grassroots levels.

Section B Individual research enquiry

Introduction

This section of the examination is your opportunity to carry out work in an area that interests you personally. It is work which is inspired and guided by your teacher but allows you to demonstrate that you can research a particular topic either in the field, or through desk-based research in the library and on the internet.

You have the opportunity to opt for one of ten themes listed in Table 2, each of which will have a topic area for the year when you take your A2 examinations (whether the exam is in January or June, the topic area will be the same). You have to research only ONE topic area. You will be examined on the topic area that you researched.

Table 2 Themes and exemplar topic areas

Theme	Topic area 2010	Other topic areas (examples)
Geography of Crime	The distribution of crime	Day and night crime
Deprivation	Deprivation in rural areas	Indicators of deprivation
Geography of Disease	The spread of a human disease	Water and disease
Environmental Psychology	Areas of residential preference	Children's perceptions of space
Leisure and Recreation	Leisure and recreation and urban regeneration	Impact of leisure on a rural area
Microclimates	Urban microclimates	Woodland microclimates
Atmospheric and Water Pollution	River pollution	Pollution and distance from the source of pollution
Geography of Retailing	Changing retailing in rural areas	Retailing and urban regeneration
Rivers	River sediments	Urban streams
Small-Scale Ecosystems	Succession	The nature of a plagioclimax

You have to select one topic area and carry out your research under a title within the area that you and your teacher agree is suitable. It is important that your research is of high quality because the examination question carries 40 UMS, or 10% of your total A-level marks.

The examination will take the form of a two-part question on your selected topic area. Part (a) will be a generic question that examines your understanding of the enquiry approach (10 marks) and part (b) will examine an aspect of your own research (15 marks).

Preparing for Section B
The structure of the individual research enquiry
Table 3 gives an indication of the timetable you should be working to. Much will depend on whether your school or college has entered you for a January or a June examination.

Table 3 Individual research enquiry timetable

January entry	Timetable	June entry	Timetable
Teacher will introduce the research themes for year	June–July at end of AS	Teacher will introduce the research themes for the year	June–July at end of AS
Confirm what topic you have selected	July before term ends	Confirm what topic you have selected	September or December
Do the research	August and/or September Year 13	Do the research	October–December Year 13 or January–February Year 13
Data processing, structuring report and writing up	October–November Year 13	Data processing, structuring report and writing up	February–April Year 13
Have your work internally assessed to make sure that it gives you a chance to answer questions about it	December	Have your work internally assessed to make sure that it gives you a chance to answer questions about it	April–May
Revise your research to enable you to answer questions	First weeks in January	Revise your research to enable you to answer questions	Late May
Sit examination	Late January	Sit examination	Early June

Pre-decision research
This is a key stage in your work. In deciding what topic area you wish to research, ask yourself the following questions:
- *What are my interests and what other A2 subjects link to the list of research themes?* For instance, if you are studying psychology, Environmental Psychology might overlap with some of your studies. Alternatively, the Geography of Crime, and Deprivation, may interest sociology students. Chemistry students might be attracted by Atmospheric and Water Pollution.
- *Which theme interests me?* If your class size is small, expect there to be restrictions on your choices.

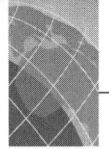

- *Have I discussed my preferred theme with the teacher and can the teacher support it?* Some departments will restrict the choice of themes for logistical reasons.
- *Do I have a potential topic of research on the selected theme?*
- *Have I found or been given background reading on the theme?* This should include articles and textbooks, as well as internet searches.
- *Are there others in the class researching similar topics?* If there are, be prepared for the teacher to expect you to carry out some shared group background research and data collection tasks.
- *Have I set myself a timetable for my research?* Things cannot be left until the last minute. The timescale is tight. Remember that research is 90% perspiration and 10% inspiration!
- Finally, *do I understand the route/sequence of enquiry?*

The route/sequence of enquiry
Read and research the literature.
1. Decide on a research topic that is either an area of investigation or a question that you wish to answer.
2. Gain the information and data from primary and secondary sources that will enable you to answer the research topic that you have chosen.
3. Understand what the information that you have gathered actually says about the topic that you are researching. How accurate is that information? Is it biased? If so, why?
4. Organise the information and data so that they answer your research aims.
5. Draw tentative conclusions from your results, which you will be able to use in the examination.
6. Evaluate each stage of your research so that you can answer questions about your sequence of enquiry in the examination.
7. Evaluate your evidence and conclusions in relation to what you read at the outset and what you have found. This should include altering the topic or the question that you asked so that others could gain a better understanding of the topic in the future.

Planning, background research and topic title setting
All good background research involves reading around the subject and solid preparatory work.
- One starting point could be undertaking internet research using the topic area title. In using the web, make sure that you know where the information came from and how valid the source is, as well as how biased it may be. Companies and organisations with reputations to maintain might structure information so as to be seen in a favourable light. The Royal Geographical Society website has a link to *Geography Now* (**www.rgs.org/now**), which has articles on topics such as 'putting poverty in its place', an ideal background for work on deprivation, or 'restoring urban rivers', potentially providing a good idea for a researcher on Rivers. Urban-based research on topics to do with deprivation, retailing, environmental

psychology and crime could make use of Google Streetview, which has video footage of streets in selected cities in the UK and abroad. This lets you take virtual walks that enable you to record information and view street-level photographs. In 2009 the system covered London, Swansea, Cardiff, Birmingham, Manchester, Liverpool, Sheffield, Leeds, York, Newcastle, Bristol, Southampton, Oxford, Cambridge, Norwich, Birmingham and Nottingham.
- Read a textbook on the topic. For instance, *Managing Ecosystems* (A. Kidd, Hodder & Stoughton, 1999) could be a start for work on small-scale ecosystems. The Geographical Association's *Discovering Cities* series provides ideas on the themes of deprivation, retailing and leisure in cities such as Bristol, Liverpool, Manchester, Portsmouth, Sheffield and Nottingham.
- Consult the journals in your institution's geography department. The following are useful sources of information:
  - *Geography Review*. For instance, the May 2004 issue had an article titled 'Investigating coastal sand dunes' — ideal for the Small-Scale Ecosystems theme.
  - *Topic Eye*. A volume of the 2009–10 series concentrated on health and development.
  - *The Geographical*. For example, the February 2009 edition had an article by P. Thomas, editor of *The Ecologist*, titled 'The trouble with travel' — potential inspiration for Leisure and Recreation research.
- If your study involves fieldwork, what equipment do you need? Can the fieldwork be completed by you on your own or would you need assistance from friends and classmates? Do you know how to use the equipment? Does the school permit you to use the equipment without staff supervision?
- If the topic requires fieldwork, select and visit the site(s) for your study.
- Prepare any datasheets that you may need to record findings. If you are going to use questionnaires, prepare the questions and pilot them on your teachers and colleagues. What type of sampling will you use? For the examination you will need to know the merits and drawbacks of the different types of sampling.
- What are the risks involved? You should assess these and your school should help you with this. If it is a topic involving Rivers or Small-Scale Ecosystems you may need to check the weather forecast; if studying tidal rivers or coastal salt marshes, check the tide times.
- Decide when you are going to do the fieldwork or when you are going to spend time on your research.
- Always ensure you have permission to enter any property or area.
- Make the topic straightforward and not too long or complicated. It must be manageable. Table 4 illustrates some straightforward topics.
- Make sure that you allocate sufficient time for:
  - data collection
  - revisiting sites or sources to improve rigour
  - writing up your research

  Create a 'time budget' with your teacher for the work.

Table 4 Potential topics for your individual research enquiry

Potential topic	Research theme
Is there a relationship between the amount and quality of green space and residential preference?	Environmental Psychology
Is there a relationship between building density and temperature in an urban area?	Microclimates
Does retailing provision vary according to the size of rural settlement?	Geography of Retailing
How important is the role of water in the transmission of malaria?	Geography of Disease
Patterns of drink-related crime and policing	Geography of Crime
Variations in river water quality	Atmospheric and Water Pollution
Contrasts in shoreline vegetation patterns	Small-Scale Ecosystems

Of these topics, several would involve fieldwork and primary data collection, while the fourth topic mentioned is dependent on good secondary sources.

Collecting the information: data collection methods

Do you know from where you are collecting the information? Do you know why you are collecting the information? What part of the topic is it intended for? Can you describe your methods of gaining information and data, and justify your use of these methods?

Presenting the information and writing the report

Once you have your data you will need to process and interpret the findings. If several of your colleagues are doing related topics, you could meet up, either with or without the tutor, to exchange information. Such a 'seminar' format will enable everybody to gain the maximum amount of understanding from the research and will probably provide you with ideas that you had not previously considered. Expect the tutor to guide you but not to provide you with all the answers.

It often pays dividends to graph, map or statistically process your information. An annotated location map for the study is essential because in the examination you will have to say what you researched and where you did the research. This is especially important in the case of primary-data and field-based studies. The Geographical Association has two relevant texts that you should consult: *Methods of Analysis of Fieldwork Data* and *Methods of Presenting Fieldwork Data*.

What kinds of diagrams should you consider?

- **Graphs**. Line graphs, histograms, pie charts (but do not overdo these), scatter graphs, frequency curves, long sections, cross-sections, and triangular graphs.
- **Maps**. Locational, choropleth, isoline, flow, distribution, and relevant published maps, such as Goad plans for shopping centres.
- **Photographs**. These can include a wide range of images, including satellite imagery, oblique and vertical air photos, Google Earth and Google Streetview

images, or your own photographs. The Panoramio online photo-sharing community (**www.panoramio.com**) provides another source of good images.
- **Statistical techniques**. Statistical and test data may be appropriate for your study. Mean, median, mode, standard deviation, scaling and weighting of data, conflict matrices, Spearman Rank, chi-squared and interpreted bipolar analyses are all useful. You should understand the sampling methods that you used to gain data.

Interpreting the information
Once the data have been satisfactorily collected, draw tentative conclusions from the data and assess the validity of the conclusions. This evaluative phase is essential because questions will be asked on this aspect of the report.

Writing up the report
A written report is not compulsory but it has the benefit of drawing together all the elements of your research investigation. By writing up a report of approximately 2,000–2,500 words, you will have produced both a completed study and, most importantly, a revision document. The report will provide the source of knowledge, understanding, application and skills used by the examiner to award you a mark (Table 5).

Knowing about your study is very important, whereas analysing, interpreting and evaluating geographical information, issues and viewpoints, and the ability to relate and apply your study to the broader field of geography, will gain you credit.

Skills form the third element of the marks and include assessment of your ability to investigate questions and issues, reach conclusions and communicate your findings in the examination by using a variety of methods and techniques as well as modern information technologies (including geographical information systems, where appropriate). Communicating your findings is very important because coherent written prose is expected in the examination and it will be easier to write well if you have completed the sequence of enquiry and revised from your own enquiry.

Table 5 How the marks are allocated for both parts of Section B

	Knowledge and understanding	Application	Skills	Total
Unit G3 Section B Total mark	9	6	10	25
Part (a)	4	2	4	10
Part (b)	5	4	6	15

Preparing for the examination
Let your teacher assess your work because he or she will be able to guide your revision on the topic towards the examination. The potential questions on the topic area will relate to both the methodology of the research study (part a) and the findings of your own research (part b). The examination assumes that you have completed individual research/investigative work, which may have included fieldwork. It assumes

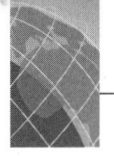

that you can identify and analyse the connections between the different aspects of geography, and analyse and synthesise geographical information in a variety of forms and from a range of sources. Crucially, the examiner expects you to be able to critically reflect on and evaluate the strengths and limitations of your enquiry at all stages: in the planning, data-collection, data-presentation, results, conclusions, and final topic re-evaluation stages.

Knowledge and understanding

To gain marks for knowledge and understanding, which comprise 50% of the marks in part (a) and 46% of those in part (b), you should be able to demonstrate in your answers that you:
- have an understanding of the key concepts that apply to your theme
- know the meaning of terms that are relevant to your study
- understand the processes that are operating in your study
- understand the interrelationship between factors in your study
- can explain where you undertook your study, how long it took and what conclusions you reached
- understand the techniques of data collection that you used and the techniques that you used to interpret the information
- know about other studies within your topic area

Application

You will gain credit if you can show in your answers that you know how to analyse, interpret and, above all, evaluate the information that you have researched. Critical reflection on and evaluation of the potential of your approach to the topic, and the limitations of your approach and methods, will demonstrate your ability to apply yourself to the topic. When researching the human themes you will also have the opportunity to gain credit for understanding people's viewpoints.

Skills

The final section of the marks will focus on the skills that you demonstrate in your answers — primarily the skill of carrying out research. Communicating your research findings in an examination requires the skill of writing in prose. Questions may also test your understanding of the data collection and processing skills used during your investigation.

In the examination

All the questions in Section B will examine the same topics no matter which research theme you studied.

The questions in part (a) will focus on aspects of the sequence or route of enquiry. You may be asked about aspects of data collection; for example: *Discuss the relative importance of primary and secondary data in an investigation into the distribution of crime* (an essentially identical title will appear for every research theme — no matter whether it is, for example, Geography of Disease, or Rivers). Other questions on data collection could focus on how you avoided bias, methods of sampling, or a critical evaluation of data collection techniques.

WJEC Unit 3

It is essential that you state the title of your investigation at the start of your answer to part (b) of the question. In this part you will always be asked to write about one aspect of your individual study.

Here are some questions that could be set using *The route or sequence of enquiry* stages outlined on page 68.

1. To what extent did your literature-based research influence the topic that you selected for your personal research?
2. How did you plan your personal research investigation?
2. Critically evaluate the planning that you undertook for your personal research investigation.
3. Evaluate the primary or secondary sources that you used in your personal research into…
3. Describe and explain the sources that you used in your research investigation.
4. Explain how you used the data that you collected in your personal research to investigate your chosen topic.
4. How did you avoid bias in either your own primary data collection or the secondary sources that you used?
5. How did you organise the data collected to answer your topic's aims?
5. Critically evaluate the data collection used in your personal research.
6. What were the conclusions that you drew from your research investigation?
6. Evaluate the conclusions that you drew from your research investigation. How certain can you be that they were valid?
8. Did the conclusions of your research investigation match those in the literature that you read before you started? If not, why not?

Many of these questions are **evaluative**, i.e. you are expected to discuss both the good and less-effective aspects of your work.

Remember
The individual research enquiry is intended to be an exercise in which you *learn for yourself with teacher guidance*. It is based on the route or sequence of enquiry (see page 68). Whatever you do during the research investigation, always maintain a critical appreciation of data and any potential bias. The investigation is small-scale so do not become overambitious. Consult your teacher on a regular basis and seek guidance. Do not do the work by merely copying from the internet — this will not help you to answer the examination questions. Remember also that your teacher will not know exactly what questions will be set in the examination, so do not rely on his/her predictions for your own revision.

Questions & Answers

This section contains some typical questions for Unit G3, with examples from Section A: Contemporary Themes and Section B: Individual Research Enquiry. All questions should be allocated the same amount of time (45 minutes).

Examiner's comments

The **examiner's comments** can be identified by the symbol ℮. A final examiner summary is also provided, which indicates the level and the grade the answer would have gained. All comments are designed to highlight the strengths and weaknesses of the candidate's answer. The answers contain typical errors, such as irrelevance, lack of focus on the question's wording, and a lack or indeed absence of case-study examples, and also illustrate the qualities needed to gain grades A or C. Spelling or grammatical errors are highlighted in bold.

Question 1

Section A Contemporary themes

Climatic hazards

This question is based on Theme 3: Climatic Hazards, Key Questions 1.5 and 1.6

Examine the factors that influence the strength of impact and the levels of risk associated with hurricanes. (25 marks)

C-grade answer

The factors that influence the severity of impact from a hurricane can be divided into two main categories. The first is physical and the second is human.*

Hurricane Katrina is an example of how even an MEDC (More Economically Developed Country) can be **devistated** by a natural hazard if the strategies to protect **them** are not used effectively.

> A plan has not been included and the candidate has restricted the answer to one event rather than looking at a range of factors to illustrate the points. It would be better to list the factors rather than mention 'physical and human', so this is a low-level introduction. The asterisk (*) here indicates to the reader that there is material added at the end. It is a good idea to review your answer for gaps, as has been done here. But the extra sentence here merely stresses that this essay may have a narrow focus. 'Devistated' is an unfortunate slip of spelling.

The physical influence around New Orleans made Katrina far more dangerous. The city is surrounded by water on three sides. This is by the Mississippi River, which is **it's** delta at this part of the country, the Gulf of Mexico and Lake Pontchartrain. To make this worse, the city is 80% below sea level by 6 to 8 feet.

To try and protect the city, levees were built but they were old and only designed to withstand the force of a category 3 hurricane.

FEMA, which is a branch of the US government that helps during large scale disasters, was aware of the danger. A training exercise called hurricane Pam was run a year before Katrina, and illustrated how New Orleans would flood if the levees were broken.

> Three physical factors affecting the impact of Katrina are stated but factors applicable to other hurricanes have not been considered. One of the factors, government, is actually human rather than physical. There is a slip of grammar in 'it's' rather than 'its' and a single-sentence paragraph, in a rather simple style of writing.

A2 Geography

question 1

The human factors are mainly a lack of preparedness and that the area had a high population density, making evacuation difficult. An evacuation order was put in place before hurricane Katrina reached New Orleans in an attempt to save people of the city. The problem was that 25% of the population of New Orleans was below the poverty line, and had no transport to evacuate. Busses were marshalled to use in the evacuation but were never actually deployed.

FEMA had gathered emergency supplies nearby as well, but it took them several days before any of this reached the people in need.**

The death toll is estimated to be 1600.

Hurricane Stan shows how, when a disaster hits a Less Economically Developed Country (LEDC) it is more dangerous than in an MEDC.

Stan was a category 1 hurricane when it hit Guatemala in 2008. There were no defences or management plans in place.

The sheer amount of rainfall caused landslides which covered entire villages. The services tried to dig into the mud, but there was no organised rescue operation, so the landslides were declared mass graves.

Hurricane Stan was only a category 1 and killed 1400 people while Katrina was a category 4 and had a similar death toll.

> The ideas here need to be grouped into meaningful paragraphs rather than single-sentence paragraphs. There are several points made that refer indirectly to risks but they are not developed. The strength of impact is also covered slightly at a tangent. This all suggests that the candidate has prepared this answer but has not done enough to adjust his/her knowledge to meet the requirements of the question. More material was added at the end, marked by the double asterisk (**), which was better organised.

Development of the **effected** country is a key factor in the level of believed risk of hazard, as this can help to reduce the number of people who die, but the main issue is the physical influences on an area. If New Orleans was not below sea level or surrounded by water it would not have flooded when Hurricane Katrina hit, indicating that physical factors are important to the level of danger in a climatic hazard.

> This paragraph is the conclusion. It does conclude that level of development is important, but then repeats some information. Also, the spelling 'affected' is required here, not 'effected'.
>
> The two written sections below are the extra paragraphs denoted by the asterisks in the text. This approach results in an essay that may lose some marks because of its lack of overall coherence.

*Hurricane Katrina made landfall in New Orleans in August 2005. It had reached category 5, but had weakened slightly before hitting the city.

WJEC Unit 3

** Many people who were unable to evacuate were instructed to take shelter in the stadium. The people inside were safe during the storm, even as the wind caused major damage to the roof. When the flooding occurred, the people in the stadium were trapped and there were issues with food and water supplies and crime.

> Overall this essay is hampered by a lack of clarity of expression and depth of support for the valid points which are being made. The stated criteria for Level 3 on the mark scheme are: 'Knowledge and understanding present but some points may be partial and lack of exemplar support. Language is in need of more complex use of geographical arguments.' Applying this suggests this essay is marginal and may just gain a C grade.

Question 2

Globalisation

This question is based on Theme 5: Globalisation, Key Questions 1.4 and 1.5

Assess the advantages and disadvantages for MEDCs of the globalisation of economic activity. (25 marks)

A-grade answer

Globalisation has many advantages and disadvantages for multinational companies in MEDCs, for workers in MEDCs and for governments in MEDCs. Globalisation is the growing interconnectedness and interdependence of the modern world. Within MEDCs, there have been consequences of this change in economic activity as a result of globalisation.

> This introduction gives a basic definition together with some indication of what the answer may contain. The ordering and repetition of phrases with 'in MEDCs' in the first sentence are not ideal but there is a good attempt to address the question.

The movement of multinational companies away from regions of the UK has resulted in mass unemployment. TNCs move away for a number of reasons. They go in search of cheaper labour and try to reduce costs. They attempt to use economies of scale by increasing production overseas to reduce the costs of transport. The new international division of labour has resulted in the movement to the periphery of workers who are paid less. This shift, also as a result of improved and cheaper transport by air or ship and improved communication over the internet, has resulted in the loss of jobs in the UK. Swansea, once a town envied for its advanced copper, zinc and steel industry, now has none left. This movement has resulted in environmental issues in the area.

> This is a good paragraph, which develops the reasons for some aspects of global shift and what it does for MEDCs. At the end it provides an example of the deindustrialisation effect of globalisation. The student uses relevant terms, although more examples might help to demonstrate that the terms are fully understood.

The environmental problems caused by heavy industry in the Tawe valley were huge. Spoil was as high as 18 m, towering over derelict factories. The R Tawe was polluted, with very low dissolved oxygen contents, making it impossible for life to survive. Leaching from spoil went into the water supply, including heavy metals such as lead and mercury. For over 60 years the site was too polluted to grow any plants and the site was an **isor**. These problems had consequences for Swansea. In order to improve environmental conditions, Swansea introduced the 5 parks scheme with the aim of redeveloping the Lower Tawe valley. The parks, called maritime park, city park, riverside park, leisure park and enterprise park, included parkland, shopping centres and retail outlets, 1500 flats in SA1 and light industry. The parks greatly improved the appearance of the region, and it now goes from strength to strength. However, there are still issues.

> The student does see deindustrialisation as a result of global shift discussed in the previous paragraph. In fact, the environmental effects were really there before globalisation, so some of the material here is less relevant. However, the loss of old industries has led to new activities which can be interpreted as being the consequence of global shift. Note 'isor', which should be 'eyesore'.

There are still high levels of unemployment, particularly among men in the area of Bon-y-Maen and Landore. Here, when men were made unemployed by the loss of heavy industry like steel manufacture, there were not enough jobs available for them. This caused resentment among many communities, particularly of the government at the time. However, new jobs have been created in these regions.

> The process of job losses and their consequences is correctly stated, although the case study is not the best.

The cost of globalisation can work in two ways. While MEDCs with TNCs move to LEDCs in search of cheaper labour, LEDC TNCs move to the UK to get closer to their markets. Products like televisions and electrical components have resulted in some new jobs in the region. Since the deindustrialisation of South Wales, newer companies have moved in. Many of these companies include food processing plants and jobs mostly in clerical work were given to women. Women began to earn more money and, with that, their status improved. This social change brought about new change in the area.

> The theoretical points all apply to globalisation's effects on an area but the case needs some real examples.

In Teeside & North East England UK steel have reduced the number of jobs available. Of over 60 steel and iron manufacturing plants in the UK in the 1970s, only 3 remain, one of which is at Redcar near Middlesbrough. The North East is one of the most deprived regions of the UK. The increased unemployment caused an increase in the crime rate as more young men get involved in gangs. Furthermore the social structure of the region changed, with an outflow of young males who moved usually down south in search of employment. As a result the community spirit has been lost.

There have been political benefits as a result of globalisation. The increased cooperation and trade has brought about investment in MEDCs like the UK. Furthermore, many Transnational Companies, like Royal Dutch Shell based in the Netherlands and the UK, have used their profits to counter environmental damage as a result of petrol emissions.

A2 Geography

> 🖉 This paragraph is less strong although it recognises political benefits. However, it drifts into environmental effects. At this point the marker will appreciate that the candidate may be rushed.

The migration of large TNCs has resulted in cheaper goods for many in the UK. Shoes that used to cost £10 can now be imported and sold for less than £3 in large supermarkets. The outsourcing of goods has meant that families and individuals in the UK can spend more on holidays, as they have larger expendable incomes and can spend more money in UK shops and businesses. This can have a positive multiplier effect, however negative multiplier effects result in the loss of employment, and can have long and far reaching consequences.

> 🖉 A relevant point has been made with support. The final sentence shows a command of terminology.

The change from secondary employment in manufacture to tertiary, quaternary and quinary employment in service industries, finance and research and development has had positives. London, Tokyo and New York are all world cities within MEDCs. They have large populations of varied origins. They have superb transport links with the rest of the world. Many TNCs choose to locate headquarters in Europe, in America and in Asia. This provides high-post jobs in management and research. This positive economic consequence is more expectable in the many larger cities, and there is more political control and influence over LEDCs for MEDCs and TNCs.

> 🖉 Good points are being made here. The candidate shows a good appreciation of the role of the service sector in globalisation and has supported his/her arguments.

Globalisation in MEDCs has had both positive and negative economic, social, political and environmental effects. From assessing the positive and negative effects, it is clear that while there have been significant improvements in the environment, there have also been losses in employment.

> 🖉 This is a very long essay to have been written in 45 minutes. It uses the language of globalisation throughout and shows good knowledge of the impacts, even if some of those stated are possibly not actually attributable to globalisation. However, the principles are correct. The essay is well organised and does offer a range of supporting evidence. Sometimes the language is rather repetitive and this is probably because the candidate has tried to include too much information. The essay has structure and good paragraphing, all of which enables the essay to gain an A grade, but probably not an A*.

Question 3

Coastal landforms

This question is based on Theme 2(b): Coastal Landforms and their Management, Key Question 1.5

Assess some of the methods used to manage coastal environments. (25 marks)

C-grade answer

There are many different management schemes available, although some systems are not suitable for areas which have particularly strong waves that are very erosive.

Some schemes are lighter management schemes which are best suited to areas that are not harshly affected by the **seas** hydraulic action, some of these management schemes are dune regeneration — this is when the **duns** are replanted so that the dunes have more stability. In order for this to be a success **fence** need to put up to stop humans from causing destruction to the dunes and trampling the plants, burrowing animals are also a big problem, e.g. Oxwich Bay where burrowing animals have weakened sand dunes. Also a big problem is trying to reduce the number of rabbits on sand dunes, this will also help the dunes to regenerate, this is a cheap way of managing erosion although it only works in less harsh environments.

Beach reprofiling is also a soft form of management and involves making the gradient of the beach steeper so that the beach absorbs the full impact of the waves. This often is more of a short term answer to the problem and doesn't make a huge impact.

Offshore reefs using **recycle** materials such as tyres **is** another soft management option. This involves putting large numbers of recycled tyres at the bottom of the sea just off the coast. The aim is to try and slow down the force of the wave and thus control the impact of the wave on the coast. In theory this sounds like a good idea although it is largely untested and the environmental impacts are not yet fully known.

Some harder forms of management which are used in more extreme cases are e.g. sea walls which are large structures built to protect the land from the sea. Swansea Bay has a large sea wall as in the 1980s many houses used to get flooded in stormy conditions. The main problems with sea walls are they are very expensive to build (around £5 million) and also expensive to maintain, up **too** £3000 a year per metre! Although they do give people security and peace of mind and also can increase the value of houses in the area by protecting them. Although this is not to say that the sea wall can't be breached by the sea.

Rip Raps are widely used as they are relatively cheap, they don't disturb the natural look of an area (they fit in) and in many cases they have been largely successful, this is when large pieces of rock are placed in front of the areas that are being battered by the sea's hydraulic action, often they are placed at the foot of a cliff or a headland.

Cliff regenerating is another hard management scheme. This is when a cliff face is regraded at an angle so it has a steady slope, this steady slope takes the **waves** force better than a straight cliff face therefore decreasing erosion, this option is unpopular with farmers who often have to sacrifice land to be regraded and completely changes the look of an area, although it is a longer term option than a quick fix.

Christchurch bay has Gabions which are large metal cages filled with stones. At Christchurch bay they have a bed of gabions at the top of the beach to stop flood water breaching the Bay and going into the town. This option is relatively cheap it isn't very unpleasing to look at and **its** quite effective, although in times of big swell and large storms the sheer force of the waves can easily move these large cages, which stops their effectiveness. Groynes can also be used, these are wood or metal posts which gather sand in order to stop it being eroded away. This method is cheap and in many areas has worked well, although in many cases it has starved other beaches along the coast of sand, so it might improve one area but it may **effect** others.

Revetments are slots of wood or metal which are placed at the top of the beach. Revetments have air space gaps in them which absorb the force of the wave when it hits. Revetments are cheaper than some of the other options e.g. sea walls and can be **affective**, although during a storm they won't be as **affective**.

In conclusion I feel that the softer options of management are better, as they are more environmentally friendly and they let nature take its course but I understand in some circumstances a tougher option is needed e.g. sea walls, when **peoples** homes and livelihoods are at risk, obviously the softer option will not work, so I think it depends on the individual circumstances.

> This is a very long essay which would have benefited from more planning. It has an introduction and a conclusion, which give it some structure. It also examines several types of coastal management, grouped as soft and hard schemes. It does assess the schemes, mainly by economic criteria (cost and property values). Where it falls down is that its quality of language is poor, with overlong sentences and poor punctuation, and problems such as confusion of 'affect' and 'effect'. The essay also lacks supporting examples other than providing the names of three areas, so there is no detail provided on the schemes. The marker will not have been convinced that the student understood marine processes and this is partly because of the quality of the language.
>
> This answer will gain a grade C for its coverage of the range of management schemes and the fact that it did assess them. In terms of the Level 4 criteria, it contains 'some critical awareness. Evaluation more patchy'. It has the following characteristics of Level 3: 'Knowledge and understanding present but some points may be partial and lack exemplar support. Mainly uses text or taught examples of variable quality.' And in terms of Level 2 criteria: 'Language is variable and slips occur'. On balance, all this places the essay in Level 3.

Question 4

Development

This question is based on Theme 4: Development, Key Question 1.4

Discuss some of the recent changes in patterns of world development. (25 marks)

A-grade answer

Patterns of world development are forever changing and geographers must forever strive to find new ways to measure the development process which are appropriate to the way that we live our lives. World development has changed for two main reasons. Firstly because of the way we assess development and secondly because of actual examples of ongoing development processes. In the past development was always seen in economic terms and, as such, the same countries came out as the most developed because of economic influence they possessed. There are now issues which come into development which determine how we regard individual countries.

One important way that the concept of development is changing and so the pattern of world development also, is the status of women in society. Women are now incredibly important to the way we assess development. Some of the OPEC countries have amassed large wealth and proportions of their population are immensely rich. However the extent to which we regard them as developed is influenced by the fact that many women are treated like second class citizens. Their opportunities are significantly lower than in other countries with similar advanced wealth because of the lack of female education and freedom. Female emancipation is an important aspect in the recent changes in how we assess development. In Britain we are one of the most developed nations in the world yet women make up only 4.8% of the managers across the country and only 2.8% of the people who sit on the boards of major UK companies. If Britain is to continue to see itself as superior in its level of development, it is important that we take into account the status of women in society. The status of women can very much be a guide to how well a country is using its wealth and the extent of its development.

The importance of human suffering should also never be forgotten in the way that we see patterns of world development. Despite China's financial muscle and burgeoning industrial supremacy, it lags behind in current world development because of the way that it treats its ordinary people. In China a lack of freedom and security affects the extent to which we can see it as developed. Oppressive state control in the media and political thought is a constant problem, a person may be imprisoned for political disagreement with the government and communist doctrine. Restrictive family policies have meant that there is a problematic gender ratio with a greater number of men than women, which in turn has given rise to the sex trafficking of vulnerable young women and forced marriages.

The treatment of indigenous people is also important to how we see world development. Yet again China is let down by the way it treats the people of Tibet, who have complete lack of political or religious freedom. Military rule is enforced and the security of ordinary people is another issue.

When discussing recent changes in the patterns of world development it is impossible not to mention the impact of sustainable development. This, unlike some other new measures of development, can be measured quantitatively to some extent by examining a country's CO_2 emissions and the percentage of energy produced through sustainable means. This form of measuring development will certainly change world development indicators into the near future. A country's ability to meet the needs of today's generations without compromising those of the future ones is very important. Some countries like the USA may be deemed underachievers in terms of sustainable development in terms of their reliance on oil instead of new, greener forms of energy. Recent changes have seen countries with sufficient financial capital begin to focus on renewable power and take a technocentric view of sustainable development issues.

Problems of world development have also been changed by ongoing development in the development process. An ideal example of this is South Korea, now a prime example of an Asian Tiger Economy. Only 50 years ago South Korea was underdeveloped, however through the influence of import substitution and US cash this East Asian country has become a major player. The South Korean government has invested heavily in large Korean companies known as **Cheebols** who have built up strong reputations on a global stage. Car manufacture is the second biggest in the world at present in South Korea.

South Korea's growth has been reinforced by the government pumping money into education and training. The modern South Korea has high aspirations and it has now achieved a society of consumers. South Korea can now be considered a newly industrialised country and it promises to keep developing, thus illustrating that patterns of world development are forever changing.

Through a combination of the way that we assess development and changes in the ongoing development process in places such as South Korea we are seeing changing patterns in world development. To ensure the actual nature of development it is essential to accept changes to the way that we measure it.

> The Level 5 criteria state: 'Very good knowledge and understanding used critically. An ability to evaluate arguments. Good, possibly original examples. A clear, coherent essay which is grammatically correct.' This essay achieves all of those points and more. It has an original idea about gender and development (an idea which may have come from studying Unit G2). It has good support and displays knowledge of the South Korean development programme (although the large conglomerates are known as *chaebols*). It is structured with an introduction and a conclusion (although this is perhaps a little rushed). All this points to a grade A*.

Question 5

Section B Individual research enquiry

Environmental psychology

(a) Outline how information may be collected in an investigation into residential preferences. (10 marks)

A-grade answer

For my investigation the aim was to determine what influences residential preferences, so I split this up into social, economic and environmental factors. I chose to focus on the market town of ..., particularly looking at the contrasting residential areas of the C19th terraced housing of ... and the modern urban suburbs of ...

Firstly, I wrote a perception study including social, economic and environmental factors which people rated 1–5 for their importance when choosing where to live. This rating gave a weighting to the different factors to investigate so that I could see whether factors which people consider important actually influence the desirability of a house, best determined by price.

Thus, I got secondary data from the environment agency on flood risk, consisting of a map showing where 50 or 100 year floods would reach. This proved largely inconclusive as only a small section of ... would suffer even 100 year floods. Other environmental **data was the traffic flow** which I conducted for 15 minutes on major through routes of each estate. To make this more useful than raw numbers though, I compared it with Ministry of Transport figures for saturation levels based on passenger car units. At the same time, noise levels were taken, tallying when the decibel meter rose above 70 or 80 dB.

For economic data, I mapped house prices for three areas [named in actual answer] using figures from estate agents, although there will be bias here as estate agents are trying to make as much money as possible from houses so will produce inflated values **on** houses for sale at the moment. Also from ... County Council website I found figures for income and socio-economic groups such as 'affluent greys' and 'moderate means'. However, the Constabulary were more helpful as they broke figures down by different wards in ... rather than the town as a whole. They also publish crime figures whereas newspaper reports focus too much on individual crimes rather than a local overview to be of much use.

Lastly, I got age and occupation figures from an online census. In conclusion I collected primary, secondary and secondary-internet data for social, economic and environmental factors from a range of sources.

question 5

☑ This is based on the theme of Environmental Psychology. The answer gives the sources of the data and discusses how the data were collected. The response outlines a range of methods of data collection although some, such as looking at individual preferences, might have been missed for the investigation named by the student. Nevertheless, there are enough sources of data mentioned. The quality of language is slightly variable but even so this answer would lead to a mark in the top level (although not full marks). This is a Level 3 response and would gain a grade A.

Question 6

Rivers

(a) **Outline how information may be collected in an investigation into river sediments.** (10 marks)

C-grade answer

In an investigation of river sediments, random sampling can be used to choose the sites to be investigated along the river. It is necessary to investigate around ten sites to gather the information needed for the investigation.

Once the sites have been chosen you need to collect the data. To collect this you must measure the width, depth, velocity, bedload and sediment load. To measure the width you measure from each bankside. To get the full width, you need to measure from the top of each bank, where the water would be after heavy rainfall. To measure the depth, you need a metre rule and a tape measure at 0.5 m. Then at every 0.5 m you measure the depth with a metre rule recording each time until you get to the other side.

[student's diagram went here]

To measure the velocity, it can be either done with a float and measuring out a 10 metre stretch along the river and timing how long it takes for the float to go the 10 metres and repeating 3 times. Or you can use a flowmeter, which is placed in the river, in a central location to obtain an average reading. The flow meter allows for a direct reading by recording the numbers of counts over one minute from which the river's speed can be calculated by using a conversion chart. The readings were repeated three times.

To measure the bedload, you use a random area sampling and collect 20 samples of the bedload. Then for each measure the Y axis of the rock and also the Cailleux roundness index. To do this you choose the pointiest end and measure it with the Cailleux index. Also you can categorise the shape with the Powers Shape Index. With this you can tell whether the rock is angular or rounded.

To measure the sediment load you need a plastic bottle with two tubes: one out of the river and one facing the flow. You hold the bottle in the middle of the river between the bed and the top and keep it there until the bottle has filled up. Then label the bottle.

> This is based on the Rivers theme. The student has mentioned the sampling techniques used, but not elaborated on their choice. The answer is competent but rather generalised, lacking development of points and specific detail, particularly with respect to the aim of the investigation.
>
> This is a Level 2 answer and would gain a grade C.

Geography of disease

(b) Summarise the main conclusions of your personal research into the spread of a disease and discuss how these conclusions support your initial aims. State the title of your investigation. (15 marks)

A-grade answer

Title An investigation into the spread of the HIV virus and more specifically, socio-economic drivers behind the spread of the virus.

From the investigation it was concluded that development (in the shape of the country's HDI score) does play a part in determining how prevalent the virus is in a particular country — those countries with a high prevalence such as Swaziland (26.1% prevalence) and Lesotho (23.2% prevalence) scored relatively poorly on the HDI scale (0.49 and 0.497 respectively). The fact that other countries within the region of Sub-Saharan Africa, such as Angola, were placed even lower on the HDI scale (0.445) and still had a relatively low prevalence (2.1%) in comparison to their neighbours would indicate that correlation between development and HIV prevalence was not as strong as expected. As a result of this further socio-economic indicators were used in order to draw conclusions regarding drivers of the epidemic. The results of the investigation showed a clear link existed between a country's fertility rate and HIV prevalence; the fertility rate of Mozambique is 5.18 and its HIV prevalence 12.5% whereas Algeria whose prevalence was just 0.1 has a fertility rate of 1.79. It is likely that this apparent link is because of the nature of the HIV virus, which can be spread from a pregnant mother to her unborn child. It was also evident from the data that the religion of the population in a particular country also plays a part in determining how quickly HIV spreads in the country. Countries with high numbers of Islamic residents such as Algeria, Morocco and Sudan consistently had low prevalence of HIV (0.1, 0.1 and 1.4% respectively), whereas countries with high numbers of Christians, in particular Catholics, e.g. Lesotho, tended to have higher prevalence (23.2%).

An expected link between wealth and HIV prevalence was tenuous. Botswana, which by African standards is a wealthy country (its GDP per capita (PPP) is $13,900), has an incredibly high prevalence (23.9%). On the other hand the Democratic Republic of the Congo, one of the poorest countries on the planet (its GDP per capita (PPP) is $300), has a fairly low HIV prevalence by Sub-Saharan African standards (4.2%).

In conclusion, less developed African nations do tend to suffer worse from HIV than more developed ones, but there are certainly other more prominent socio-economic drivers of the HIV epidemic in Africa.

> This is based on the Geography of Disease theme. This answer provides much detailed support from the research findings, concerning the factors influencing the incidence of HIV in Africa. The answer summarises conclusions, yet does

WJEC Unit 3

not directly refer to the second part of the question. The candidate has assumed that the aims are covered in the title that was given at the start of the answer.

The response is worth a mark on the A/B borderline because it has excellent command of the findings but it does not make clear how the findings relate to the initial aims. This response would be placed at the bottom of Level 4 or the top of Level 3, and might just gain an A grade.

Leisure and recreation

(b) Summarise the main conclusions of your personal research into leisure, recreation and urban regeneration and discuss how these conclusions support your initial aims. State the title of your investigation. (15 marks)

C-grade answer

'As You Get Closer to ..., Urban Regeneration, Leisure and Recreation are More Apparent'

My personal research followed the hypothesis above, I was investigating the link between leisure and regeneration in relation to urban regeneration, after collecting all of the information in question (a) I looked at **me** results and concluded that the hypothesis was very accurate.

My pedestrian count showed that the number of people in the regenerated area was considerably higher than in the non regenerated area, this linked directly to the fact that regeneration brought with it facilities such as canoeing, sailing and many more along with cafes, restaurants and shops, meaning the reason for the difference in the amount of people is largely due to these shops and facilities.

My litter count showed that there was less litter in the regenerated area although there were more people, meaning the council **were** taking good care of maintaining this area regularly. In the other areas there seemed to be far more litter however very few people about, being almost overshadowed by the regenerated area.

The car registrations in the regenerated area showed that the majority of cars were new and expensive whereas in the other areas they were quite old and used cars, **this supports that** the more affluent people owning newer, more expensive cars have more disposable income, allowing them to participate in leisure and recreation activities which is why they were in the regenerated area of ....

Census data showed that the people living in the regenerated area had high paying jobs or a dual income with a partner suggesting that accommodation is expensive.

The land use maps were very useful because they showed the scale of regeneration, the amount of regeneration that had been done and the number of leisure and recreation facilities. I could also plot my pedestrian count on the map showing what facilities people were close to.

Questionnaires suggested the majority of people were active and were using the facilities on the river.

My hypothesis proved to be very accurate and followed the trend of activities at ....

☒ This essay outlines the investigative techniques used and the conclusions reached. However, because there is no effective evaluation of those conclusions as required by the question, the candidate cannot access the higher marks. The last sentence provides a token evaluation. Also, the quality of the language, particularly the punctuation, is variable and detracts from the flow of the essay. Some of the results could have been critically examined to test their veracity.